Daniel Kopetzki

Exploring Hydrothermal Reactions

Daniel Kopetzki

Exploring Hydrothermal Reactions
From Prebiotic Synthesis to Green Chemistry

Südwestdeutscher Verlag für Hochschulschriften

Impressum/Imprint (nur für Deutschland/only for Germany)
Bibliografische Information der Deutschen Nationalbibliothek: Die Deutsche Nationalbibliothek verzeichnet diese Publikation in der Deutschen Nationalbibliografie; detaillierte bibliografische Daten sind im Internet über http://dnb.d-nb.de abrufbar.
Alle in diesem Buch genannten Marken und Produktnamen unterliegen warenzeichen-, marken- oder patentrechtlichem Schutz bzw. sind Warenzeichen oder eingetragene Warenzeichen der jeweiligen Inhaber. Die Wiedergabe von Marken, Produktnamen, Gebrauchsnamen, Handelsnamen, Warenbezeichnungen u.s.w. in diesem Werk berechtigt auch ohne besondere Kennzeichnung nicht zu der Annahme, dass solche Namen im Sinne der Warenzeichen- und Markenschutzgesetzgebung als frei zu betrachten wären und daher von jedermann benutzt werden dürften.

Verlag: Südwestdeutscher Verlag für Hochschulschriften GmbH & Co. KG
Dudweiler Landstr. 99, 66123 Saarbrücken, Deutschland
Telefon +49 681 37 20 271-1, Telefax +49 681 37 20 271-0
Email: info@svh-verlag.de

Approved by: Potsdam, University, Diss., 2011

Herstellung in Deutschland:
Schaltungsdienst Lange o.H.G., Berlin
Books on Demand GmbH, Norderstedt
Reha GmbH, Saarbrücken
Amazon Distribution GmbH, Leipzig
ISBN: 978-3-8381-2915-0

Imprint (only for USA, GB)
Bibliographic information published by the Deutsche Nationalbibliothek: The Deutsche Nationalbibliothek lists this publication in the Deutsche Nationalbibliografie; detailed bibliographic data are available in the Internet at http://dnb.d-nb.de.
Any brand names and product names mentioned in this book are subject to trademark, brand or patent protection and are trademarks or registered trademarks of their respective holders. The use of brand names, product names, common names, trade names, product descriptions etc. even without a particular marking in this works is in no way to be construed to mean that such names may be regarded as unrestricted in respect of trademark and brand protection legislation and could thus be used by anyone.

Publisher: Südwestdeutscher Verlag für Hochschulschriften GmbH & Co. KG
Dudweiler Landstr. 99, 66123 Saarbrücken, Germany
Phone +49 681 37 20 271-1, Fax +49 681 37 20 271-0
Email: info@svh-verlag.de

Printed in the U.S.A.
Printed in the U.K. by (see last page)
ISBN: 978-3-8381-2915-0

Copyright © 2011 by the author and Südwestdeutscher Verlag für Hochschulschriften GmbH & Co. KG and licensors
All rights reserved. Saarbrücken 2011

Contents

Summary   5

1 Introduction   7

2 Fundamentals   11
    2.1 Green Chemistry 11
       2.1.1 Current Status 11
       2.1.2 Principles of Green Chemistry 12
       2.1.3 Biorefinery 13
       2.1.4 Choice of Solvents 16
       2.1.5 Synthesis in Water 17
    2.2 Properties of Water 18
       2.2.1 Structure 18
       2.2.2 Water Anomalies 20
       2.2.3 Interaction with solutes 21
           2.2.3.1 Nonionic solutes 21
           2.2.3.2 Ionic solutes 22
    2.3 Hydrothermal Water — Physicochemical Properties 23
    2.4 Hydrothermal Synthesis 28
       2.4.1 Biomass Valorisation 28
       2.4.2 Syntheses 30
           2.4.2.1 Hydrolysis — Water as Reactant 30
           2.4.2.2 Condensation Reactions — Water as Product 32
           2.4.2.3 Potential of NCW in Organic Synthesis 33
    2.5 Outline 35

Contents

3 Experimental — 37
- 3.1 Synthesis — 37
  - 3.1.1 Batch Mode — 37
  - 3.1.2 Continuous Flow Reactor — 38
- 3.2 Gas Chromatography/Mass Spectroscopy — 40
  - 3.2.1 Gas Chromatography — 40
  - 3.2.2 Mass Spectroscopy — 42
- 3.3 Analytics of Selected Species — 43
  - 3.3.1 Identification of Unknowns by Silylation — 43
  - 3.3.2 Carbohydrate Analysis — 44
  - 3.3.3 Amino Acid Analysis — 47
  - 3.3.4 Formaldehyde Analysis — 48

4 Prebiotic Carbohydrate Synthesis — 49
- 4.1 Background — 49
- 4.2 Moderate-Temperature Reaction — 52
- 4.3 Hydrothermal Formose Reaction — 55
  - 4.3.1 Effect of Added Salt — 55
  - 4.3.2 Product Identification — 56
  - 4.3.3 Influence of Catalytically Active Species — 61
  - 4.3.4 Characteristics of the Hydrothermal Reaction — 63
  - 4.3.5 High Buffer Concentration — 66
  - 4.3.6 Carbohydrate Selectivity — 68
  - 4.3.7 Effect of Temperature — 70
  - 4.3.8 Sugar Stabilization — 70
- 4.4 Summary — 76

5 Transfer Hydrogenation of Levulinic Acid — 77
- 5.1 Background — 77
- 5.2 Results — 79
  - 5.2.1 Salt Effects — 79
  - 5.2.2 Dissociation at High Temperature — 82
  - 5.2.3 Optimization of Reaction Conditions — 84

	5.3	Summary	86
6	**Hydrothermal Biomass Valorization**		**89**
	6.1	Alkaline Digestion	89
		6.1.1 Conversion of Glucose	89
		6.1.2 Liquefaction of Wood	91
	6.2	Stability and Decomposition of Glycine	94
		6.2.1 Theory	94
		6.2.2 Results	94
	6.3	Summary	101
7	**Conclusion and Outlook**		**103**

Appendix   105

Instrumental Details 105
Chemicals 106

Table of Symbols   107

List of Figures   109

List of Tables   113

Bibliography   115

Summary

In this thesis chemical reactions under hydrothermal conditions were explored, whereby emphasis was put on green chemistry. Water at high temperature and pressure acts as a benign solvent. Motivation to work under hydrothermal conditions was well-founded in the tunability of physicochemical properties with temperature, e.g. of dielectric constant, density or ion product, which often resulted in surprising reactivity. Another cornerstone was the implementation of the principles of green chemistry. Besides the use of water as solvent, this included the employment of a sustainable feedstock and the sensible use of resources by minimizing waste and harmful intermediates and additives.

To evaluate the feasibility of hydrothermal conditions for chemical synthesis, exemplary reactions were performed. These were carried out in a continuous flow reactor, allowing for precise control of reaction conditions and kinetics measurements. In most experiments a temperature of 200 °C in combination with a pressure of 100 bar was chosen. In some cases the temperature was even raised to 300 °C.

Water in this subcritical range can also be found in nature at hydrothermal vents on the ocean floor. On the primitive earth, environments with such conditions were however present in larger numbers. Therefore we tested whether biologically important carbohydrates could be formed at high temperature from the simple, probably prebiotic precursor formaldehyde. Indeed, this *formose* reaction could be carried out successfully, although the yield was lower compared to the counterpart reaction under ambient conditions. However, striking differences regarding selectivity and necessary catalysts were observed. At moderate temperatures bases and catalytically active cations like Ca^{2+} are necessary and the main products are hexoses and pentoses, which accumulate due to their higher stability. In contrast, in high-temperature water no catalyst was necessary but a slightly alkaline solution was sufficient. Hexoses were only formed in negligible amounts, whereas pentoses and the shorter carbo-

Summary

hydrates accounted for the major fraction. Amongst the pentoses there was some preference for the formation of ribose. Even deoxy sugars could be detected in traces.

The observation that catalysts can be avoided was successfully transferred to another reaction. In a green chemistry approach platform chemicals must be produced from sustainable resources. Carbohydrates can for instance be employed as a basis. They can be transformed to levulinic acid and formic acid, which can both react via a transfer hydrogenation to the green solvent and biofuel γ-valerolactone. This second reaction usually requires catalysis by Ru or Pd, which are neither sustainable nor low-priced. Under hydrothermal conditions these heavy metals could be avoided and replaced by cheap salts, taking advantage of the temperature dependence of the acid dissociation constant. Simple sulfate was recognized as a temperature switchable base. With this additive high yield could be achieved by simultaneous prevention of waste. In contrast to conventional bases, which create salt upon neutralization, a temperature switchable base becomes neutral again when cooled down and thus can be reused. This adds another sustainable feature to the high atom economy of the presented hydrothermal synthesis.

In a last study complex decomposition pathways of biomass were investigated. Gas chromatography in conjunction with mass spectroscopy has proven to be a powerful tool for the identification of unknowns. It was observed that several acids were formed when carbohydrates were treated with bases at high temperature. This procedure was also applied to digest wood. Afterwards it was possible to fermentate the solution and a good yield of methane was obtained. This has to be regarded in the light of the fact that wood practically cannot be used as a feedstock in a biogas factory. Thus the hydrothermal pretreatment is an efficient means to employ such materials as well. Also the reaction network of the hydrothermal decomposition of glycine was investigated using isotope-labeled compounds as comparison for the unambiguous identification of unknowns. This refined analysis allowed the identification of several new molecules and pathways, not yet described in literature.

In summary several advantages could be taken from synthesis in high-temperature water. Many catalysts, absolutely necessary under ambient conditions, could either be completely avoided or replaced by cheap, sustainable alternatives. In this respect water is not only a green solvent, but helps to prevent waste and preserves resources.

Chapter 1

Introduction

The recent oil spill in the Gulf of Mexico has once again shown quite plainly the dangers of the exploitation of fossil resources. This incident is just one in a series of many other severe accidents. Risks from the usage of oil are, however, not limited to such acute events, but persist on a much longer timescale. Combustion of oil and of products produced thereof results in the exhaustion of large quantities of carbon dioxide. Atmospheric levels of this green house gas have increased dramatically over the last century, and nobody can estimate with certainty how great its impact will be on the global climate.

An increased environmental consciousness has led to the development towards a cleaner chemistry. The principles of this *green chemistry* are not limited to the tapping of sustainable resources. It is rather a holistic concept, minimizing risk and pollution from the point of manufacture to the ultimate disposal of the products. This includes the use of benign solvents as well as the elimination of toxic intermediates.

In this context water is gaining increasing importance. Due to the long lasting predominance of organic solvents, synthesis in water is far from being fully explored. Another reason is the sensitivity of most catalysts and reagents towards it, which requires the investigation of novel alternatives. In this thesis the focus is directed towards water under hydrothermal conditions. At a temperature above the normal boiling point, the liquid state is retained by applying high pressure. The presented experiments are carried out at temperatures up to 300 °C using pressures around 100 bar.

The goal of this work is the evaluation of hydrothermal conditions for performing

Chapter 1 Introduction

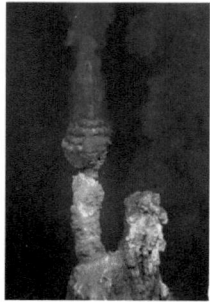

Figure 1.1: A hydrothermal vent and the striking biosphere at this environment [1].

chemical reactions, under the premise of green chemistry. The motivation to work at high temperature can partly be found in the tunability of physicochemical properties. Many characteristics of water massively change when approaching the critical point, i.e. the dielectric constant, the ion product or the density. These changes can have dramatic effects on reactivity, and it is an ambitious aim to exploit these properties. Indeed, many benefits could be drawn for the systems presented herein.

Water at sub- and even supercritical conditions can be found in nature at hydrothermal vents on the ocean floor. These biotopes accommodate a striking marine fauna amidst an otherwise hostile environment (see figure 1.1). On the primitive earth such hot habitats were present in much larger numbers. This was the motivation to study whether biologically important molecules, in particular carbohydrates, could have been formed under such conditions from the simple, possibly prebiotic precursor formaldehyde. Despite the instability of the products this reaction was indeed successful and furthermore revealed crucial differences compared to the counterpart reaction under ambient conditions. Surprisingly, selectivity was altered and less demanding requirements concerning catalysts were necessary.

Carbohydrates are also the major part of biomass and thus a sustainable feedstock to synthesize platform chemicals that nowadays are nearly exclusively manufactured on the basis of fossil resources. γ-Valerolactone is such a versatile molecule and can be produced from sugars. Furthermore, it is nontoxic, a green solvent and may be used as biofuel. A key step in its production from carbohydrates is the hydrogenation of levulinic acid. In this thesis we successfully performed this reduction using formic acid, a by-product of a former step. This usually requires the use of heavy metal catalysts, namely palladium or ruthenium. Addition of simple salts was explored instead, and indeed sodium sulfate surprisingly proved to be a cheap, sustainable alternative. Here the temperature dependence of the acid dissociation constant of the anion was exploited, rendering certain salts to temperature switchable bases. This concept might be expanded to other base catalyzed reactions as well, thus minimizing waste and toxic additives.

As high-temperature chemistry is characterized by surprising reactivity, it is often necessary to identify unknown products in order to elucidate new reaction pathways. For this purpose mass spectroscopy was performed. By using isotope labeled compounds as comparison, the structure of unknowns could be identified, also when present only in traces. With this technique the complex hydrothermal reaction network of glycine was investigated. By this refined analysis novel compounds and pathways could be identified, which were not yet described in literature.

Chapter 2

Fundamentals

2.1 Green Chemistry

2.1.1 Current Status

Not least due to the availability of crude oil, chemistry has emerged into a field which now is present in all areas of life. Polymers have become a major building block for all imaginable kinds of commodities. Nearly all materials that surround us in our daily life are treated with chemicals in some way. Material prosperity and economic wealth both are based on the disposability and processing of fossil resources. This dependency bears a great danger in two respects. Firstly, usage of fossil resources will release carbon dioxide and is thus of great environmental concern. This carbon has been bound in oil or coal for millions of years and its combustion will perturb the susceptible carbon equilibrium of nature. Secondly, we are facing a shortage of this feedstock and consequently new sustainable sources of raw materials have to be tapped if we want to keep the present state of wealth and technical culture.

The enormous challenge of coping with these problems is addressed by the principles of *Green Chemistry*. It is defined as "design of chemical products and processes to reduce or eliminate the use and generation of hazardous substances" [2, 3]. This is achieved by designing novel synthesis pathways as well as using sustainable resources and techniques.

2.1.2 Principles of Green Chemistry

Approaches towards green chemistry should meet several principles, which were first summarized in 1998 by Anastas [4]. As they are the basis of the present thesis, they shall be briefly discussed in the following.

1. **Prevention.** Waste should preferentially be prevented instead of treating it afterwards.

2. **Atom Economy.** All employed chemicals should be incorporated into the final product.

3. **Less Hazardous Chemical Synthesis.** Substances that are used during synthesis should possess little or no toxicity to humans and environment.

4. **Designing Safer Chemicals.** Toxicity of products should be reduced.

5. **Safer Solvents.** Auxiliary substances and solvents should be avoided or, when necessary, be innocuous.

6. **Energy Efficiency.** Energy requirements for chemical processes should be as low as possible.

7. **Renewable Feedstock.** A sustainable feedstock is superior to fossil resources.

8. **Reduce Derivatives.** Derivatization, like the use of protecting groups, must be minimized to save additional chemicals.

9. **Catalysis.** Catalytic reagents are superior to stoichiometric ones.

10. **Design for Degradation.** Products should be degradable and not persist in the environment.

11. **Real-Time Analysis.** Analytical methods should be constructed for real-time and in-process monitoring.

12. **Accident Prevention.** All types of hazards should be addressed and avoided.

Applying these recommendations is challenging, as alternative feedstocks, solvents and synthetic pathways have to be explored [5]. These principles not only serve academic interest but are already being incorporated into industrial chemical synthesis. Processes that are environmentally benign can in fact be economically more profitable as well, not to mention the marketing advantages. Examples of such processes include the production of pesticides as well as pharmaceuticals [2, 6, 7].

2.1.3 Biorefinery

One of the most ambitious goals of green chemistry is the replacement of fossil resources by renewable materials. This process is challenging as sustainable compounds are very different from fossil resources. Consequently a new chemistry is required to deal with such a feedstock. Renewable resources can be considered as all those produced from biomass. Even when taking into account combustion, they can be regarded as CO_2 neutral, as they are part of the biological carbon cycle.

A biorefinery is a plant that produces useful raw materials, analogous to a traditional petroleum refinery, however, using renewable resources as feedstock. The main components of biomass are carbohydrates, lignin and smaller quantities of amino acids and oils [8–10]. Contrary to petroleum, the oxygen content of these compounds is higher, requiring a novel processing technique. However, not only purely chemical transformations can be applied. Biotechnology, the use of enzymes and fermentation play an important role, too.

Carbohydrates are of particular interest for biorefineries, as they are the major constituent of biomass. Several important raw materials, which can be produced from them, have been identified. These include ethanol, furfural, hydroxymethylfurfural, lactic acid, succinic acid, levulinic acid, xylitol and sorbitol [11]. A majority of them can be obtained by fermentation via bacteria or yeasts. Furfural, hydroxymethylfurfural or levulinic acid on the other hand can be synthesized by chemical pathways.

Lactic acid is the product of glucose fermentation using specialized bacteria. It can be polymerized via a ring opening polymerization of the dilactide [12]. Polylactic acid is a biodegradable polymer. Nevertheless, it is quite resistant and shows properties similar or superior to widespread polymers like polyethylene terephthalate, polyethylene or polystyrene [13]. These traditional polymers are of concern, as they are very

Chapter 2 Fundamentals

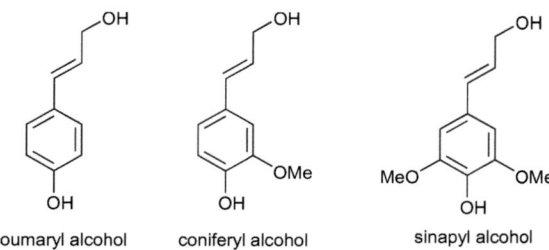

Figure 2.1: The three main building blocks of lignin.

persistent in the environment. Manufacturing beverage bottles, plastic bags or food packaging with polylactic acid would solve some waste problems. However, the costs of this polymer still exceed those of petroleum based ones, though with shortage of fossil resources we can expect that to change in the future.

Carbohydrates are fairly easy to process. Unfortunately this does not apply for lignin, the second most abundant component in wood. Lignin is a highly cross-linked polymer derived from coumaryl, coniferyl and sinapyl alcohol (figure 2.1) [14]. These monomers are polymerized by peroxidase enzymes to large networks. Lignin is responsible for the stability of wood. With a fraction of about 30% in wood [15], lignin is the major byproduct of the paper industry. Nowadays only a minor fraction is used, e.g. for the production of vanillin [16]. Due to the lack of convenient methods to split natural lignin into monomers, the majority is simply burned. Recently a processable thermoplastic material, *arboform*, was produced from lignin waste of pulp industry and plant fibres [17].

Fats are another natural source. Biosynthesis involves the esterification of glycerol with fatty acids. Transesterification with methanol is used in huge scales to produce biodiesel [18, 19]. Fatty acids are an interesting building block with high synthetic significance [20]. However, glycerol accumulates as waste in large amounts. This demonstrates another challenge of green chemistry, the complete use of the feedstock and possible by-products, which is currently faced by the design of new selective catalysts [21–23]. This task is hampered due to the high degree of functionalization of biomass.

Besides raw materials for chemical synthesis, biorefineries are also important re-

2.1 Green Chemistry

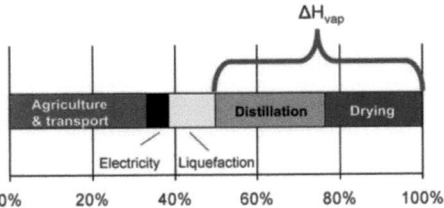

Figure 2.2: Distribution of energy requirements in corn-grain ethanol production, taken from [25].

garding fuels [24]. In fact, nearly 95% of total products from conventional refineries are employed as transportation fuels and for energy generation. Biomolecules are too polar to be used directly. Even plant oil usually requires an additional transesterification step. So chemical modification aiming for the removal of oxygen is necessary. In this thesis the synthesis of a nontoxic biofuel based on carbohydrates will be presented. Another means for deoxygenation is pyrolysis. Even lignin can be converted to a mixture of oxygen-poor hydrocarbons [15]. However, it is chemically disadvantageous to break down the complete molecular structure. Instead it would be of higher value to selectively manipulate structural units in the naturally occuring compounds.

Besides the advantages of biorefineries compared to conventional plants based on fossil resources, there are also some concerns. One is related to the feedstock. Both major ways to produce biofuel, the fermentation of starch to ethanol and the transesterification of fats to biodiesel, rely on agricultural products. They thus compete with food production. As a consequence the price of some groceries has increased. This problem is addressed by focusing on alternative feedstock, e.g. waste, low intensity energy crops or lignin. Another important point is the efficiency. For most processes the net energy balance is if at all only marginally favourable. The energy inputs for the ethanol production from corn grain are shown in figure 2.2. Most of the energy is consumed for manufacturing. In this process distillation and drying, both related to the high vaporization energy of water, make up the major fraction. The elimination of these inefficient steps is a big challenge.

15

2.1.4 Choice of Solvents

Most chemical reactions are performed in solvents. These are a major source of waste as recycling is often energy consuming. Some reactions can also proceed without solvents, which of course is one of the favoured processing ways. Unfortunately, this does not apply to most reactions.

Here, the green chemistry approach led to several possible innovations. One is the design of novel green solvents. In this thesis the synthesis of such a solvent, γ-valerolactone, will be presented. Contrary to conventional solvents, it is not harmful to the environment but even biodegradable. A lot of effort is being put into ionic liquids as well [26]. These are salts that are either liquid at room temperature or exhibit a low melting point. Their advantage is the very low vapour pressure and low flammability. Because both cation and anion can be varied, the term "designer solvents" is attributed to ionic liquids. Reactions that can be performed in these solvents are highly versatile. However, it can be problematic to isolate the product afterwards. One possibility is selective extraction, for example with supercritical CO_2 [27].

Supercritical (sc) fluids themselves can be used as solvents. As their properties depend strongly on temperature and pressure, they are tunable. ScCO_2 has solvent properties comparable to light hydrocarbons and is one of the cheapest and most widespread supercritical fluids [28]. As the critical temperature is only 31 °C, it can be processed quite easily, although high pressure equipment is necessary. The advantage of CO_2 is the easy separation of solvent and product after reaction. Releasing the pressure will degas the system, and the solvent is removed. Supercritical CO_2 has found a widespread application in industry, e.g. for the decaffeination of tea and coffee beans and for dry cleaning. Here it successfully replaced toxic organic solvents like perchloroethylene. Although CO_2 incontrovertibly is a greenhouse gas, it can be taken out of the air and thus is part of the natural carbon cycle.

Water can be considered the most green solvent. It is ubiquitous, cheap and does not generate any hazards. Therefore, it is particularly interesting for large scale applications. Especially for the mostly polar biomass, water is a good solvent. Furthermore, many enzymes, which in most instances rely on the presence of water, can be employed. When nonpolar products are obtained, separation form water can be carried out easily by decantation instead of extraction.

2.1.5 Synthesis in Water

Despite the important role of conventional solvents in organic synthesis, reactions in water are continuously explored, and many examples are already known [29, 30]. Water cannot only replace organic solvents, in some cases even an enhancement of reactivity is observed. This was first demonstrated for the Diels-Alder reaction of cyclopentadiene with butenone [31]. The rate in water is much higher than in isooctane or methanol, which is attributed to the hydrophobic effect. Nonpolar species tend to aggregate in water, which in the case of the Diels-Alder reaction brings the two reactants into proximity. Additionally, the endo vs. exo selectivity was improved. Similarly, other pericyclic reactions work well in water.

Besides the hydrophobic effect, the high dielectric constant of water can also accelerate reactions involving an ionic transition state. Examples are nucleophilic substitutions that run according to a S_N1 mechanism [32]. But also nucleophilic additions, e.g. the common aldol reactions, can be performed in water. This is an important class of reactions for the establishment of C-C bonds.

For basically all reactions usually performed in organic solvents, an example in water can be demonstrated. This includes oxidations, reductions or transition metal mediated transformations. However, many novel reagents are required as most conventional ones are sensitive to moisture or simply insoluble. This also involves the design of new catalysts. For instance in recent years many water tolerable Lewis acids have been identified. Traditionally employed $AlCl_3$ for example hydrolyzes in water and then becomes inactive. In contrast, some rare-earth metal triflates and other salts retain their activity [33, 34].

Nature provides many chiral compounds, most of them water soluble. This allows their use as cheap, green auxiliaries and catalysts for assymetric reactions in water [35]. Amino acids can serve as chiral ligands, for example in the enantioselective reduction of acetophenone [36]. Cr^{2+}, complexed by different amino acids, was employed as reducing agent for this step. Another means to induce stereoselectivity is the use of chiral auxiliaries. Sugars were used to introduce both water solubility and a source of chirality into a diene [37]. Diels-Alder reaction and subsequent removal

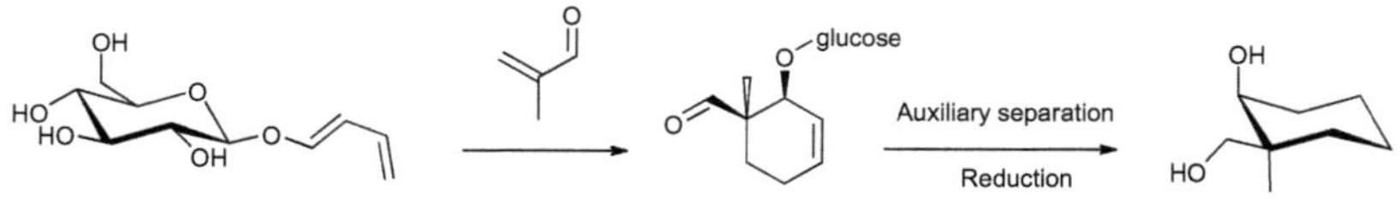

Figure 2.3: Glucose as chiral auxiliary in a Diels-Alder reaction.

of the carbohydrate yielded the desired product with high endo vs. exo selctivity and modest diastereomeric excess (see figure 2.3).

With growing applications of biotechnology, reactions in water will become more important. Most enzymes require aqueous conditions for high activity. Also when proteins have to be modified, often an organic solvent cannot be used, as biomolecules can rapidly undergo denaturation. Here, water is not an option but a necessity. Because of the huge advantages of water regarding green chemistry, it was used as solvent for all presented syntheses in this thesis. Hence, the properties of this remarkable molecule shall now be elucidated in more detail.

2.2 Properties of Water

2.2.1 Structure

Liquids in general can be divided into two groups. Solvents of the first group are relatively unstructured and described as "regular" liquids. They are only held together by weak van der Waals interactions. They are packed closely and each molecule has about 10–11 nearest neighbours [38]. The second are structured liquids, which are held together by much stronger forces. They can have fewer nearest neighbours. One water molecule can for instance form up to four hydrogen bonds. This is realized in hexagonal ice, the natural most abundant modification of solid water. One water molecule with its surrounding four neighbours is visualized in figure 2.4. Such a structured network has very low entropy and displays many cavities. Small molecules, for example methane, can occupy these spaces in so called clathrates. Large quantities of this methane hydrate are indeed found on the sea floor.

The triple point of water is at around 0.01 °C. Here, liquid water is in equilibrium with ice and vapour. Many modifications of ice are known, so there are actually

2.2 Properties of Water

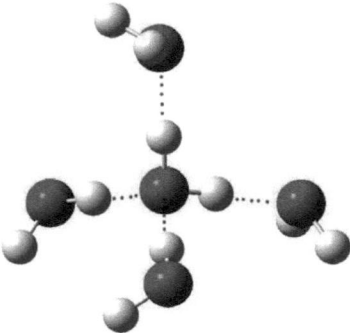

Figure 2.4: One water molecule can form four hydrogen bonds (calculated with *Gaussian 03*).

other triple points as well. The critical point of water is at 374 °C and 221 bar. Even supercritical water can be found on earth. Volcanic activity in the deep sea leads to the formation of so called hydrothermal vents. Compared to most other chemicals, the water phase diagram shows some astonishing peculiarities (figure 2.5) [39]. One example is the density, which is higher for liquid than for solid water. Such a contraction upon melting is rare among substances. This behaviour can be rationalized by the network of hydrogen bonds in ice. They strongly keep together the molecules, but are also responsible for the spacious assembly. Upon melting this well ordered structure partially collapses, resulting in a higher density. Under ambient pressure and temperature, each molecule has in the liquid state 4.4 nearest neighbours on average [40], but still the tetrahedral coordination is more or less retained.

Because of the strong intermolecular interactions, the presence of water clusters has been proposed. However, there is much controversy, and many contradictory values are given in literature, ranging from small aggregates with only few molecules to bigger ones with more than hundred. The difficulties of explaining water arise from the 3D-hydrogen bond network. A simple hard core repulsion potential cannot describe molecular interactions but rather the intermolecular and directional hydrogen bonding. The average strength between two water molecules is 20 kJ mol^{-1}. This

Chapter 2 Fundamentals

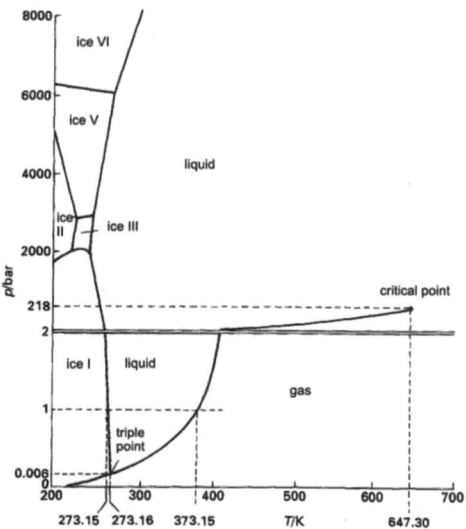

Figure 2.5: Phase diagram of water [39].

is about ten times kT, the thermal fluctuation energy at room temperature. The strongly directed (ordered) interactions are manifested in the molar entropies. At 273 K and 1 atm they are 41 J K^{-1} mol^{-1} for ice and 63.2 J K^{-1} mol^{-1} for liquid water, respectively. These values are quite close, whereas in the gas phase the molar entropy is 188 J K^{-1} mol^{-1}. Despite these strong interactions, the hydrogen bond network is highly dynamic with reorientation times in the ps range [41].

2.2.2 Water Anomalies

Water is far from being a simple liquid. It is laborious to describe water with theoretical models, and many properties are not yet understood [42, 43]. The anomalies of water are crucial for evolution and existence of life. The density increase upon melting has already been explained. Even upon further heating of liquid water it still increases, reaching a maximum at 4 °C [44]. At a sufficiently high pressure, this phenomenon disappears.

The strong hydrogen bonding in liquid water and the tetrahedral configuration are the key factors for the anomalies of water [45]. In contrast to NH_3 or HF, which also form hydrogen bonds, water is liquid at room temperature. One water molecule has two hydrogen atoms and two lone pairs respectively, whereas this number is unequal for NH_3 and HF. Furthermore, the tetrahedral configuration of water creates prolonged 3D structures.

Anomalies that can be explained by the strong molecular interactions are the large heat capacity and the heat of vaporization. Also the high boiling point can be rationalized. Other properties are more difficult to understand, like the aforementioned density maximum, the minimum in isothermal compressibility around 46 °C or the minimum in specific heat capacity. The anomalies can be classified according to structural, dynamic and thermodynamic properties [46]. With increasing temperature they disappear. In this context the structural anomalies are more resistant to temperature, while the thermodynamic ones are the first to vanish.

Simulations are also used to calculate water properties. Particularly successful have been approaches assuming a second critical point of water. Although this is speculative, some experimental indications exist. When water is cooled rapidly an amorphous solid can be obtained. There exist two different states, low-density amorphous water (LDA) and high-density amorpous water (HDA) [47]. This has been taken as an indication that correspondingly two states of liquid water exist, one with low and the other with high density. However, the critical point is assumed to be at very high pressure and low temperature [48]. As water cannot be supercooled to this condition, the point is experimentally inaccessible. Even if the existence of a second critical point could be clearly demonstrated, it would not explain the anomalies of water, but rather add another.

2.2.3 Interaction with solutes

2.2.3.1 Nonionic solutes

Polar molecules are readily dissolved in water. These compounds can form strong hydrogen bonds with water. However, hydrophobic molecules also show a slight solubility in water. The solubility of water in oil is even more significant. The term hydrophobic implies that nonpolar compounds would dislike water. Nevertheless, the solubilization of aliphatic hydrocarbons is an exothermic process. There is a favourable interaction of nonpolar compounds with water by van der Waals attraction. The enthalpy is, however, overcompensated by a highly unfavourable entropy. The decreased entropy upon solubilization can be interpreted as an increase in order. For many years it was believed that water in the hydrophobic hydration shell shows increased hydrogen bonding and greater tetrahedrality. Because this resembles crystalline water, this assumption is called "iceberg theory". New experiments and simulations have shown that there is hardly any increased order. The loss in entropy is probably a result of the excluded volume effect. Water molecules cannot occupy the space where the solute is situated. Also the reduced motional freedom of water around the solute contributes to the decrease in entropy.

If we regard infinitely diluted solutions, we can observe a so called hydrophobic hydration shell around the solute. At higher concentrations these shells start to overlap and bring the solute molecules closer. This process, the so called hydrophobic effect, is entropy driven, because some water is released. This can accelerate organic reactions in water by merging the reactants. This effect was already mentioned for the Diels-Alder reaction (page 17).

2.2.3.2 Ionic solutes

Contrary to hydrophobic solutes, most salts show high solubility in water. Water in the first hydration shell is bound very strongly. Hydrophilic anions form strong hydrogen bonds, whereas cations interact with the negative charge of the water oxygen atom. The second hydration shell, with a larger number of water molecules than in the first, is bound less strongly. In most cases ion-pair formation is negligible at room temperature.

Concerning the nature of ions, two types can be distinguished. This was first studied in the context of the influence of salts on protein solubility. Some salts enhance solubility, whereas others lead to precipitation. In general, the effect of anions dominates over that of cations. The ions can be ordered according to their degree of influence in the *Hofmeister series* [49], shown in table 2.1. The first group are the kosmotropic ions. The interaction of the ion with water is stronger than

2.3 Hydrothermal Water — Physicochemical Properties

Table 2.1: The Hofmeister series

	kosmotropic				$\Rightarrow$			*chaotropic*	
Anions:	F^-	PO_4^{3-}	SO_4^{2-}	Cl^-	NO_3^-	Br^-	I^-	ClO_4^-	SCN^-
Cations:	Me_4N^+	NH_4^+	K^+	Na^+	Cs^+	Li^+	Mg^{2+}	Ca^{2+}	Ba^{2+}

the intermolecular attraction of water itself. These ions are called salting-out ions. They decrease the solubility of nonpolar compounds in water and strengthen the hydrophobic effect. Furthermore, they stabilize proteins against denaturation. The reverse behaviour is found for chaotropes, the salting-in ions. On the one hand they increase the solubility of proteins, but on the other hand proteins are also destabilized due to higher vulnerability towards denaturation.

The influence of salts on biomolecules was related to the structure of water. Kosmotropes were believed to enhance the ordering of water, whereas chaotropes are structure-breaking species. New spectroscopic evidence shows that these salts do not have any influence on the structure of bulk water and its hydrogen network. Only rotation of water in the first solvation shell is restricted [50]. The influence of ions on biomolecules now seems to be purely an effect of direct ion-protein interaction and of the interaction of ions with the hydration shell of the biomolecule.

2.3 Hydrothermal Water — Physicochemical Properties

In the preceding paragraphs some properties of water have been presented. However, the aforementioned anomalies vanish with increasing temperature. Water could then be considered as a "normal" liquid. This however does not decrease interest in it. In fact, some characteristics of high-temperature water make it a versatile green solvent.

The critical point of water is at 374 °C and 221 bar. The harsh conditions in supercritical water (SCW) make it difficult to perform selective reactions. Nevertheless, some examples with relatively stable compounds are known [51]. So far, the most useful applications involve reactions where no selectivity is required. This includes the gasification of biomass [52] or supercritical water oxidation (SCWO)

Chapter 2 Fundamentals

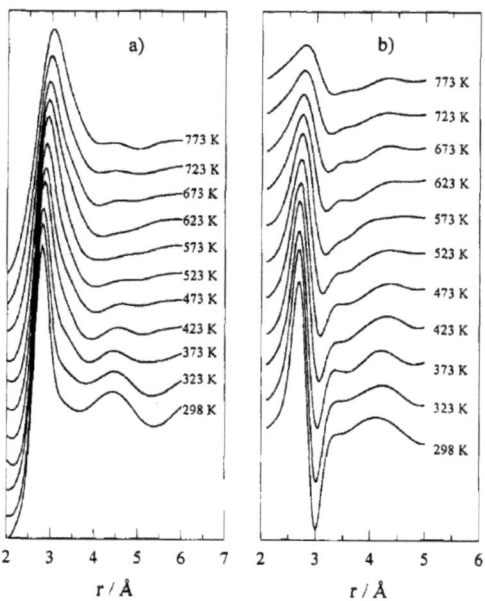

Figure 2.6: Molecular pair correlation functions of liquid and supercritical water at 1 kbar (a) and its first derivative (b), taken from [54].

[53]. The latter is used to completely destroy organics by oxygen or other oxidants dissolved in supercritical water. With respect to synthesis, subcritical conditions are more promising. Water at an intermediate temperature between ambient and supercritical conditions (around 200–300 °C) is referred to as high-temperature water (HTW), near-critical water (NCW), hot compressed or subcritical water. When ions are present, the term ionothermal can be applied.

Properties of hydrothermal water differ significantly from those of ambient water. The main reason is the structure change due to loss of hydrogen bonding. This is indicated by the pair correlation functions of liquid and supercritical water, shown in figure 2.6. The experiments were carried out at 1 kbar. The first peak around 2.8 Å corresponds to the shortest intermolecular oxygen-oxygen separation. At higher

2.3 Hydrothermal Water — Physicochemical Properties

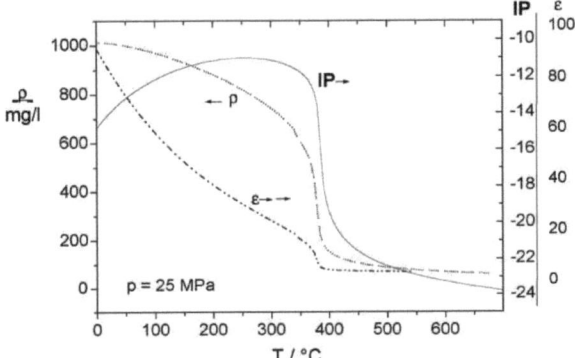

Figure 2.7: Properties of water at high temperature under 250 bar, taken from [56].

temperature it is shifted towards larger distances and the intensity decreases, whereas the second neighbour peak (4.5 Å) becomes smaller and finally disappears upon further increase in temperature. This reflects a loss both of hydrogen bonding and in tetrahedral coordination of water [55]. Compared to ambient temperature, about 55–60% of the hydrogen-bonding network is lost at 300 °C. However, these data also indicate that even at the highest temperature studied there are still some hydrogen-bonded molecular pairs.

This weakening of the water structure influences many other properties, as shown in figure 2.7. The density ρ of water strongly decreases with increasing temperature. This becomes more pronounced when approaching the critical temperature. In the near-critical region the changes are less distinct. At saturation pressure the density decreases from 1 g cm^{-3} at 25 °C to 0.75 g cm^{-3} at 300 °C. One consequence of the reduced density is the improved transport due to an increase in diffusion. It is also a decisive factor for the dominance of ionic mechanisms in subcritical water, as opposed to free radical ones in supercritical water. In principal properties can also be adjusted by tuning the pressure. Due to the high compressibility, this is especially useful regarding supercritical conditions.

A special feature of high-temperature water are the increased acidity and basicity. At saturation pressure the ion product K_w increases by nearly three orders of mag-

Chapter 2 Fundamentals

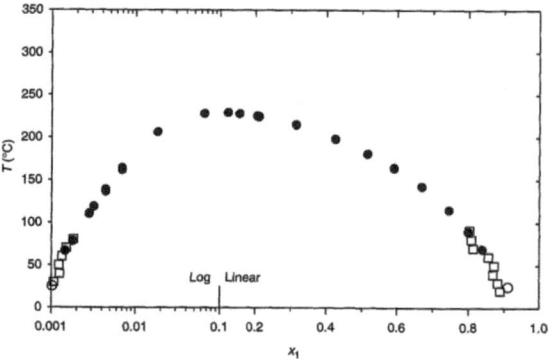

Figure 2.8: Liquid-liquid equilibria of acetophenone (component 1)/water, taken from [30].

nitude, reaching a maximum around 250 °C. The advantage is that both acid and base catalyzed reactions are accelerated in near-critical water. Regarding sustainability, this is especially important, as such reactions can be performed by simple heating, without additional acids or bases. Cooling the mixture restores the initial neutral conditions. This saves chemical neutralization steps and thus prevents waste. Under high pressures in the kbar range, K_w can be further increased up to almost 10^{-9} mol^2 L^{-2} [57]. However, due to the drastic conditions necessary this is predominantly of academic interest.

Even more notable is a change in the dielectric constant ϵ. Here, vast changes already occur in the near-critical region. The high dielectric constant of ambient water facilitates solubilization of ionic and polar species. At 300 °C and saturation pressure the dielectric constant is reduced to approximately 20. This is in the range of solvents like acetone. The reduced dielectric constant increases the solubility of nonpolar compounds, whereas it reduces the solubility of inorganic salts. Above an upper critical solution temperature (UCST) organic compounds are fully miscible with water. Nonpolar species have a high UCST, e.g. n-hexane (355 °C) or benzene (305 °C), whereas the presence of functional groups lowers the UCST. One example is acetophenone with an UCST of 228 °C, shown in figure 2.8 [30]. In the water-rich side, the solubility of acetophenone increases almost exponentially with temperature.

2.3 Hydrothermal Water — Physicochemical Properties

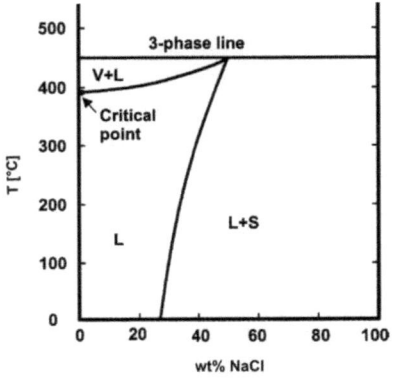

Figure 2.9: Phase diagram of NaCl/water at 250 bar, taken from [59].

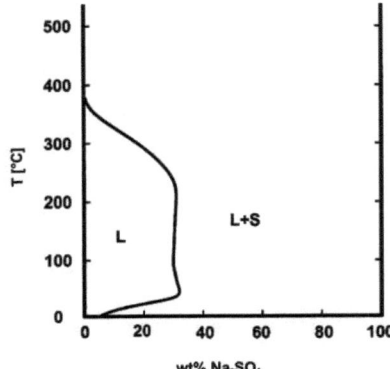

Figure 2.10: Phase diagram of Na_2SO_4/water at 250 bar, taken from [59].

This is quite a general phenomenon for nonpolar compounds. On the organic-rich side the increase in solubility with temperature is more gradual. Even at ambient temperature there is a noticeable solubility of water, as the carbonyl functionality offers some possibility for hydrogen bonding. The strong dependency of solubility on temperature, especially in the aqueous regime, offers superb possibilities for product recovery, namely phase separation simply induced by cooling.

On the other hand, care must be taken when inorganic salts are present in NCW. Because of loss of hydration water with increasing temperature, salts are primarily solubilized as ion pairs by which solubility can be drastically reduced [56]. Experimental results for various salts under hydrothermal conditions are available [58]. These show that two different types can be classified. Salts of the first group exhibit high solubility in the vicinity of the critical point. NaCl is one example, its phase diagram under a pressure of 250 bar is shown in figure 2.9. With increasing temperature its solubility becomes even higher, despite the formation of ion pairs. Above the critical point however, the solubility of salt is very low. Also, the position of the critical point is altered when solute is present. Na_2SO_4 and other salts of the second group show a different behaviour (see figure 2.10). The solubility already decreases rapidly before reaching the critical point.

Chapter 2 Fundamentals

Under supercritical conditions the solubility of inorganic salt is extremely low, with density being the decisive factor. By applying high pressures, thereby increasing density, solubility can be enhanced. As in this work reactions were carried out in NCW, the importance of checking the solubility of added salts should be emphasized. Precipitation of salts at high temperature is a commercial problem. Equipment can easily be blocked or damaged, especially when working in flow.

2.4 Hydrothermal Synthesis

2.4.1 Biomass Valorisation

High-temperature methods are especially useful when processing biomass. It has already been stated that the chemical use of agricultural products competes with food supply and is ethically questionable (see page 15). A relatively old method of processing any kind of biomass, including waste and therefore eliminating this concern, is pyrolysis [60]. In the absence of oxygen, biomass is heated to high temperatures between 400–1000 °C. Depending on the conditions, various amounts of char, bio-oil and gas are obtained. Compared to biomass, these have a lower oxygen content but a higher heating value.

An energetic disadvantage of classical pyrolysis is a preceding drying step. As most biomass is wet, this additional step can be skipped when working in water. Depending on the temperature, different products can be obtained. General trends are visualized in figure 2.11. At very high temperatures of around 600 °C biomass is converted mainly to H_2 and CO_2. When conditions are carefully chosen, the formation of char can be avoided. High yields can be obtained for most feeds, with the constraint that lignin is more difficult to process.

When the temperature is lowered (around 400 °C), selectivity is shifted towards the production of CH_4 at the expense of H_2. Usually a metal catalyst is necessary to prevent the formation of char. As a first step towards gasification, water soluble intermediates are formed from biomass. These are mainly furfurals and phenols. The mode of action of the catalyst involves the rupture of the C-C bonds and supply of O˙ and OH˙ radicals from the dissociation of water molecules. These fragments ultimately react to form methane. The methane formation must be fast enough since

2.4 Hydrothermal Synthesis

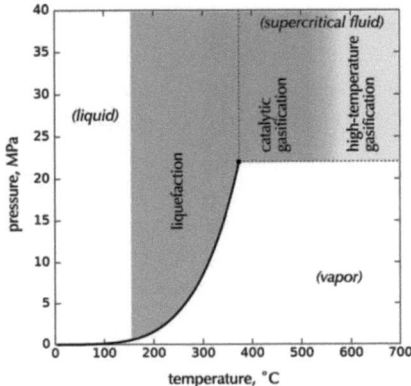

Figure 2.11: Hydrothermal processing regions and preferential products, taken from [25].

otherwise the reactive intermediates can form a stable polymeric material.

Motivated by the biological origin and formation of oil, it was suggested that biomass can be liquefied in hydrothermal water. Indeed, when heating biomass in NCW some viscous oil is formed. This process is catalyzed by bases. Usually temperatures between 280–380 °C and pressures up to 300 bar are employed. During this liquefaction, the initial oxygen content of 30–50% is reduced to 10–20% and the heating value is at least doubled [25]. Performance is much better than for flash pyrolysis, a fast pyrolysis of dry biomass. The bio-oil is water insoluble, so that separation is not an issue. Of course, its oxygen content is still higher than that of conventional petroleum, preventing traditional petrochemical processing. Furthermore, the oil is not stable but prone to polymerization. However, even difficult substrates like wood can be liquefied. A first commercial plant is already in operation. Changing World Technologies, Inc. uses wastes from a turkey farm to produce diesel oil, fertilizers and carbon [61]. Its capacity is around 100 tons per day.

At even lower temperatures (200 °C) hydrothermal carbon can be obtained from carbohydrate-rich biomass [62]. Nearly all of the initial carbon is preserved in the final product. The material is oxygen-rich and was proposed as soil conditioner and a method for CO_2 sequestration.

Chapter 2 Fundamentals

Bio-oil is a broad mixture of undefined compounds [52]. In order to synthesize commodity chemicals with high purity, it is necessary to introduce more selective reactions yielding defined compounds. Recent progress was made in the synthesis of lactic and acetic acid [63]. Glycerol, which is available in large amounts from the biodiesel production, can be converted in alkaline media to lactic acid and hydrogen with 90% yield. Carbohydrates too can yield this acid under catalysis by some metal ions, for instance Zn^{2+}. Apart from these small molecules originating from biomasss, much interest is also focused on organic synthesis. The variable properties of high-temperature water make it a versatile solvent in this respect.

2.4.2 Syntheses

2.4.2.1 Hydrolysis — Water as Reactant

In organic synthesis much care is taken to remove water. Water is the by-product of many reactions, and thus its presence shifts the equilibrium to low product yield. On the other hand the reverse reaction, namely the hydrolysis, is also of interest, for instance when protecting groups have to be removed. Examples are esters, ethers, acetals, alkyl halides, cyano groups or amides [64]. These reactions can be catalyzed by acids or bases. The increased ion product under hydrothermal conditions can thus aid hydrolysis.

The hydrolysis of methyl *tert*-butyl ether was measured at 250 bar under sub- and supercritical conditions [65]. The reaction is first order. When plotting the logarithm of the rate constant against the reciprocal temperature, a striking jump around the critical point of water is detected (figure 2.12(a)). The reaction is faster in NCW than under supercritical conditions. Kinetics at 600 °C and 300 °C are comparable. This can be rationalized by the self-dissociation of water. As mentioned before, it is increased at hydrothermal temperatures but strongly decreases when reaching the critical point. When dividing the apparent reaction rate by the proton concentration, a linear plot is obtained (figure 2.12(b)). This shows the high potential of NCW in hydrolysis reactions, as an additional acid catalyst can be avoided.

Esters are also readily hydrolyzed in neat high-temperature water. This is of technical interest concerning the workup of natural fats or polyethylene terephthalate. The latter is largely used for the production of plastic bottles. In hydrothermal water

2.4 Hydrothermal Synthesis

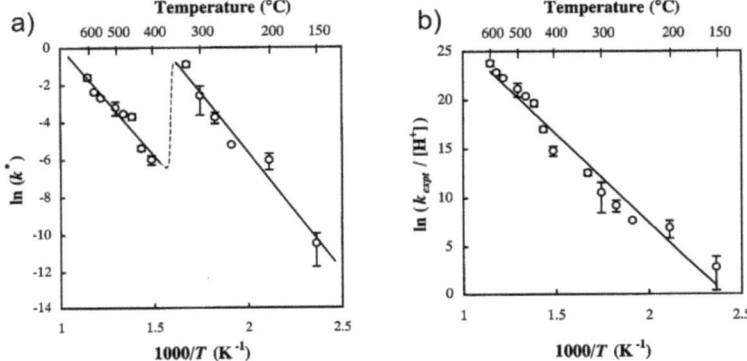

Figure 2.12: Disappearance of methyl *tert*-butyl ether; apparent rate constant (a) and pH-corrected rate constant (b), taken from [65].

it is soluble and the depolymerization towards the monomers ethylene glycol and terephthalic acid proceeds smoothly. This reaction can be catalyzed by ammonia [66]. Other widespread polymers can also be depolymerized in NCW, including polyamides like nylon. Despite the higher temperature, which means higher energy input, these reactions are economically viable, as kinetics is much accelerated.

Even reactions that do not readily occur at moderate temperatures can be efficiently performed in NCW. Nitro compounds can be hydrolyzed to the corresponding alkanes [67]. This reaction is acid catalyzed (mechanism see figure 2.13). Because of the relatively mild conditions without added acid, this reaction proceeds without further hydrolysis of aniline. When using phosphoric acid as catalyst and working under harsher supercritical conditions even aniline can be hydrolyzed. Temperature serves to adjust selectivity. This also applies to carboxylic acid groups. Below a

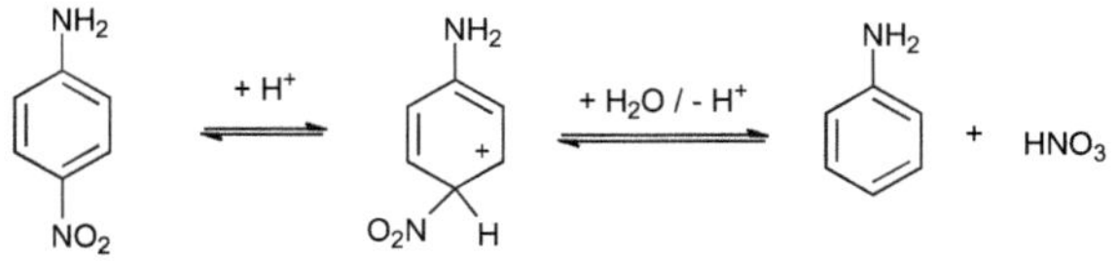

Figure 2.13: Acid-catalyzed mechanism for the hydrolysis of 4-nitroaniline.

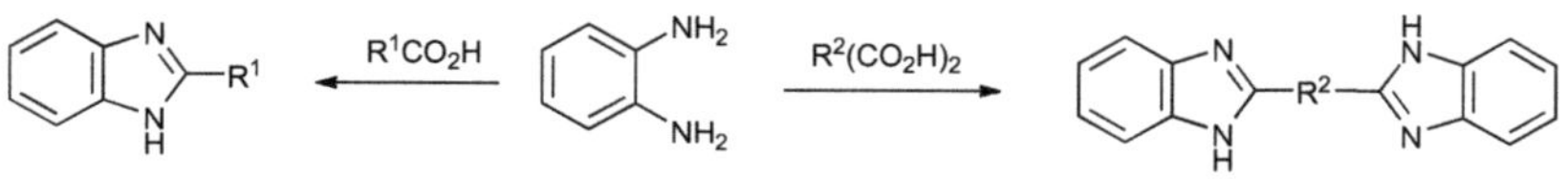

Chapter 2 Fundamentals

Figure 2.14: Synthesis of substituted benzimidazoles from 1,2-phenylenediamine.

certain temperature they stay intact, but when working at higher temperature decarboxylation takes place.

2.4.2.2 Condensation Reactions — Water as Product

Although an aqueous environment seems contradictory, elimination reactions can proceed well under hydrothermal conditions. The dehydration of an alcohol to the corresponding alkene is thermodynamically favoured at high temperature. Furthermore, this reaction is catalyzed by acids. Cyclohexanol undergoes nearly complete dehydration in NCW and also the aromatic compound 4-chloro-2-propyl benzyl alcohol was dehydrated without effects on the substituents or polymerization [68]. Of course, these elimination reactions just proceed until thermodynamic equilibrium is reached. The reverse reaction, the hydration of alkenes to alcohols, also takes place, although yields are low at high temperature.

A typical condensation reaction, the formation of benzimidazole, is shown in figure 2.14. At 350 °C in water without additional catalysts the maximum yield was 90% [69]. After a double dehydration a cyclization occurs to yield the final product. Again, no additional acid was necessary to promote the reaction.

Besides this advantage, greener reactants can also be employed in hydrothermal reactions. One important reaction to introduce alkyl chains into aromatics is the Friedel-Crafts alkylation. Usually, it is performed using halogenated alkanes. A Lewis acid such as $AlCl_3$ acts as catalyst. In the hydrothermal counterpart these harmful substances can be replaced. A simple alcohol like *tert*-butanol can serve as the alkylating agent. Through protonation and dehydration an electrophilic carbocation is formed, which easily reacts with activated aromatics like phenol. The reaction is reversible and yield is higher with increasing temperature. The main products of the hydrothermal Friedel-Crafts alkylation are ortho and para substituted compounds, which one would also expect at conventional reaction conditions. However, in water

2.4 Hydrothermal Synthesis

Figure 2.15: The Beckmann rearrangement of cyclohexanoxime yields ϵ-caprolactam.

the ratio of ortho to para can be tuned with temperature [56]. Even Friedel-Crafts acylation reactions can be performed under hydrothermal conditions. Traditionally, stoichiometric amounts of Lewis acid are necessary when working with acylchlorides as reactants. In high-temperature water substitution by carboxylic acids is possible, however, the less favourable thermodynamics results in a low yield [70].

Apart from these acid catalyzed reactions, basic catalysis also works well in NCW. At ambient conditions aldol condensations and Cannizzaro reactions only occur in very alkaline media. However, they can be performed in neat water at sufficiently high temperatures. Due to the high temperature the intermediately formed alcohol dehydrates easily. It is not surprising that the retro-aldol reaction occurs in water as well. Even glucose can be converted to glycolaldehyde with 64% yield in only 0.25 s when working in supercritical water (450 °C) at low density. The selectivity is quite high, considering that glucose can easily yield several decomposition products.

2.4.2.3 Potential of NCW in Organic Synthesis

The spectrum of possible reactions in NCW is of course not limited to hydrolysis and condensation. The synthetically very versatile Diels-Alder reaction was already mentioned. Rearrangements, like the Beckmann rearrangement (see figure 2.15), an important step in the production of nylon 6, can also be carried out in water. Addition of H_2SO_4, a typical catalyst of this reaction, is not necessary. Various oxidation, reduction and even organometallic reactions work in water. To gain some overview, this diversity is classified in figure 2.16.

Although there are many benefits of hydrothermal water, some limitations should be considered. When water is formed as product, the yield is thermodynamically limited. The high dissociation constant, although useful for catalysis, makes NCW a very aggressive medium. Expensive equipment is necessary to reduce corrosion. The harsh conditions, high temperature accompanied by high pressure, are a safety

33

Chapter 2 Fundamentals

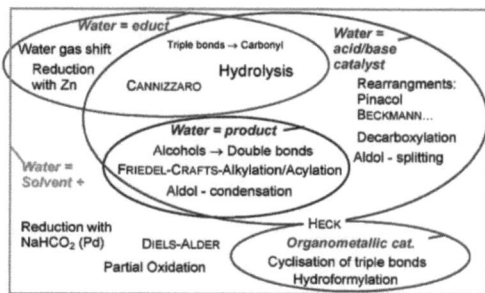

Figure 2.16: Scheme of synthesis reactions carried out in high-temperature water, taken from [56].

concern. Furthermore, the temperature must be carefully chosen as not to destroy sensitive groups. Side reactions can be an important reason for reduced selectivity. Because of different possible pathways many products can be formed [71], which is rarely desirable. The energy needed to reach high temperatures is remarkably high. However, efficient reactor design can minimize energy input. When exothermic reactions are performed the produced heat can be sufficient to maintain a high temperature. This autogenic processing even applies to supercritical conditions, e.g. in SCWO, where no external heating is necessary once the reaction is running.

Apart from these limitations, high-temperature water has proven to be a versatile green solvent offering unique properties. These are not limited to the environmental benefit of replacing organic solvents. Temperature can be tuned to precipitate products, thus removing them from equilibrium. Both acid and base catalyzed reactions can be performed in neat water. This avoids waste as subsequent neutralization is not necessary. Less expensive catalysts may be employed. Product separation can be fairly easy due to the decreased solubility of non-polar products at lower temperature. The tunability of NCW allows the control of selectivity, e.g. ortho/para or endo/exo, which is hard to achieve with conventional processing techniques. Considering the huge potential of hydrothermal synthesis, only preliminary results are available. It will be the scope of this work to contribute some insights into the feasibility of high-temperature water as reaction medium. The green approach is not only limited to the choice of this solvent, but is the basis for the selected reactions.

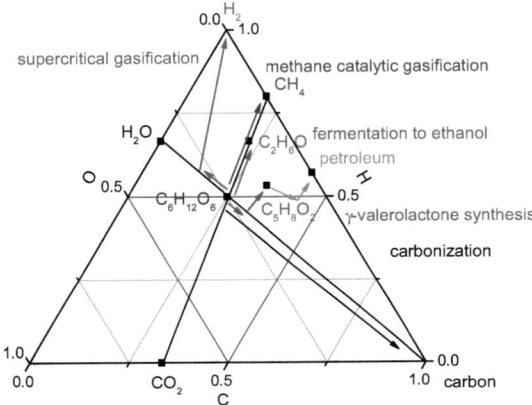

Figure 2.17: Conversion landscape of carbohydrates by various pathways through elimination or incorporation of H_2O and CO_2.

2.5 Outline

The goal of this thesis is the hydrothermal conversion of sustainable compounds to useful chemicals. Water as the most green solvent will be employed throughout. Additional chemicals will be chosen according to the principles of green chemistry. The basic starting material in most experiments are carbohydrates. Possible transformations are illustrated in a CHO-diagram using the example of hexoses (figure 2.17).

Several processes that have already been mentioned are shown. The supercritical gasification to hydrogen requires a formal uptake of water and release of CO_2. In the catalytic gasification towards methane at lower temperatures only CO_2 is lost but no water eliminated. This also applies for the biochemical fermentation to ethanol. The pathway to carbon is followed when solely water is lost. At various stages of carbonization the C-content increases. Considerable amounts of oxygen are incorporated in hydrothermal carbon produced at 200 °C, whereas anthracite coal is nearly pure carbon.

In this thesis alternative pathways are explored. In the first section the fundamental question concerning the origin of our starting material, the carbohydrates, will be elucidated. A simple, possibly prebiotic synthesis via the formose reaction using formaldehyde is presented. Furthermore, this study reveals crucial differences to the counterpart reaction under ambient conditions. Insights into the stability of carbohydrates at high temperature are gained.

The second part will deal with the synthesis of the green solvent and biofuel γ-valerolactone from carbohydrates. It involves the transfer hydrogenation of levulinic acid with formic acid as hydrogen donor. The hydrothermal synthesis does not rely on the presence of toxic heavy metal catalysts. Instead, the tunability of thermodynamic parameters with temperature is exploited, using Na_2SO_4 as a temperature switchable base. During this synthesis 1 eq CO_2 and 2 eq H_2O are eliminated. Continuation of these elimination processes would ultimately lead to the formation of petroleum (compare figure 2.17).

In the last part, the focus is directed towards complex plant material. A hydrothermal alkaline pretreatment of biomass proves to be an efficient means for increasing kinetics and yield of subsequent fermentation to biogas. Furthermore, the hydrothermal decomposition pathways of glycine are investigated. Intermediates are identified by mass spectroscopy using isotope labeled compounds, which allows the construction of an extended reaction network under hydrothermal conditions.

Chapter 3

Experimental

3.1 Synthesis

3.1.1 Batch Mode

As hydrothermal reactions run at elevated pressure, specialized equipment is necessary. The vapour pressure of water is already around 15 bar at 200 °C and increases rapidly when approaching the critical point. The easiest means to work under such conditions is to use sealable vessels, here referred to as autoclaves, that withstand these harsh conditions. Due to the high pressure, such devices are usually manufactured from metal alloys. Stainless steal bombs are most common, but other materials like hastelloy, tantalum or titanium are employed as well for more corrosive liquids.

After filling and sealing, the autoclaves can be heated, resulting in the built-up of the solvent's autogeneous pressure. External pressure adjustment is not possible for the majority of commercially available systems. Most effective heating methods comprise the immersion of the reactor in molten salt baths [66]. Here, the intended temperature can be reached in few minutes. However, this setup is inadequate for very fast reactions.

An alternative way is microwave-assisted synthesis. Glass or quartz vessels are employed, since the material must be permeable to the electromagnetic radiation. Due to the limited maximum pressure, conditions are usually restricted to the sub-critical range of water. As the solvent is heated directly and the process not limited by heat transfer through the reactor wall, this technique offers superior control of temperature and heating rate. The transparent reactor material together with the

Chapter 3 Experimental

indirect heating enable visual control of reaction progress. Microwave reactors have become very popular in organic synthesis. Working at temperatures above the ambient boiling point of the solvent remarkably speeds up reactions.

Since the introduction of microwave-assisted synthesis, the question of whether the observed increase in kinetics is solely an effect of temperature or if there exists any peculiar "microwave effect" has been debated controversially [72]. It is now generally accepted that the speed up of chemical reactions is mainly attributed to thermal effects. Considering the Arrhenius equation

$$k = a \cdot e^{-E_a/RT} \tag{3.1}$$

the rate constant k is only affected by the temperature. Neither is there a significant specific microwave effect on the preexponential factor a nor on the activation energy E_a. In any case, due to its instantaneous heating, has microwave-assisted synthesis become a widespread issue in organic chemistry.

3.1.2 Continuous Flow Reactor

Performing the reaction in standard autoclaves is only feasible when reaction times of several hours are intended, as an extended time is required to heat up the equipment to the desired temperature. This prohibits kinetic studies. Furthermore, there is normally no external control of pressure. Both problems can be overcome by using a continuous flow arrangement. This technique became popular with the development of combinatorial chemistry and the need for fast, reliable micro-scale synthesis [73]. In industry, most processes are carried out in flow. Besides the relative ease of handling small quantities in micro channels [74], flow synthesis can be automatized fairly easily. Solution can be pumped through columns, in which heterogeneous catalysts or immobilized reagents are packed. Since several of these can be combined in succession, multiple synthesis steps can conveniently be carried out in one run. Besides, such systems can be equipped with online-analysis.

With the integration of a pressure regulator, flow reactors can be feasible for hydrothermal synthesis as well. A typical setup is shown schematically in figure 3.1. This arrangement shares some common features with an HPLC, e.g. a high pressure

3.1 Synthesis

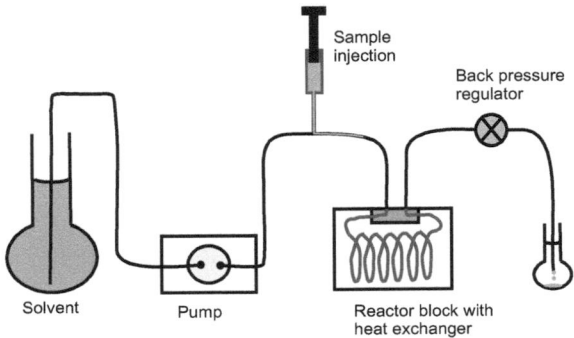

Figure 3.1: Schematic overview of a high pressure continuous flow reactor.

pump with adjustable flow rate. Solvent is flushed through thin capillaries. Due to the high corrosivity of high temperature water, in particular when salts are present, the reactor which was used for experiments presented herein was made of Hastelloy C-22. This alloy is based on nickel, chromium and molybdenum and withstands reducing and oxidizing media even at elevated temperatures. Due to the thin tubes, heat transfer rates are extraordinary high. Incoming solution is heated rapidly by a heat exchanger at the inlet of the reactor block, which also applies for the cooling process at the outlet. Flow reactors are suitable for reaction times of less than 1 s, during which even supercritical conditions can be achieved [75].

Also, mass transfer is highly accelerated in capillaries, as diffusional mixing occurs rapidly due to the small tube diameter [76]. Other advantages are the thermal management of exothermic reactions, which are controllable and safe because of the small amounts employed. One application is the destruction of explosives by supercritical water oxidation [77]. Once successful reaction conditions have been identified, upscaling is easy. Products simply have to be collected for a prolonged time without changing reaction parameters, whereas using bigger vessels in batch reactions might result in unpredictable effects.

An important device in the presented setup is the back pressure regulator. It allows to block the flow partially and thus adjust the pressure of the system. The employed setup can handle pressures in the range of 50–180 bar. Sample is not injected until the system is equilibrated with pure solvent. Due to the fast heat transfer and the fixed flow rate, the reaction time can be set with high accuracy. However, when calculating the residence time, the reduced density of high-temperature water has to be taken into account, as the flow rate of the pump refers to solvent under ambient conditions. Because of the relatively low concentration of solute, the density of pure water was used [78] and the residence time t_r calculated according to

$$t_r = \frac{V}{q} \cdot \frac{\rho(T_r, p_r)}{\rho(T^\circ, p^\circ)} \qquad (3.2)$$

with the reactor volume V, the flow rate q and $\rho(T^\circ, p^\circ)$ and $\rho(T_r, p_r)$ being the density of water under ambient and reaction conditions.

3.2 Gas Chromatography/Mass Spectroscopy

3.2.1 Gas Chromatography

Products were mostly analyzed by gas chromatography (GC). Like in all chromatographic techniques, analytes are separated based on their different affinity for a stationary and a mobile phase. As the name implies, the mobile carrier is a gas (helium or hydrogen), whereas the stationary phase can be either a solid or a liquid.

The basic setup of a GC is shown in fig. 3.2. Analytes are transferred into the gas phase by injection into a heated chamber, which is flushed with carrier gas. Thus GC is only applicable to thermally stable and volatile compounds. The vapour can then be directly transferred onto the column. This is called splitless injection and is mainly used for trace analysis. In most cases only a small fraction of the injected sample is necessary. In the split mode the remains are purged through a valve between injector and column. The small amount injected results in a narrow starting band. In the splitless mode sharp peaks are obtained by solvent focusing. Choosing a solvent with a boiling point high enough, it will condense in the first part of the column. Due to the higher affinity for the liquid solvent, analytes will be solubilized within it. As the solvent slowly evaporates, induced by an increase in oven temperature, analytes are focused to a small spot.

Separation is achieved as a result of different retention by the column, which fur-

3.2 Gas Chromatography/Mass Spectroscopy

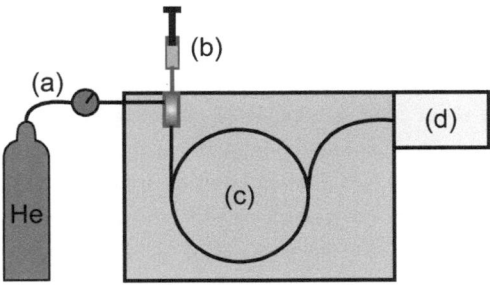

Figure 3.2: The basic setup of a GC consists of a gas supply (a), injector (b), oven and column (c), and detector (d).

thermore depends on the temperature of the oven. The first columns employed were packed with silica particles that acted as solid support coated with a liquid layer. Today so called WCOT-capillary (well coated open tubular) columns are used. These are large glass capillaries coated with a thin liquid film, often chemically bonded to the support material. The typical length is between 30–100 m. Because of the small inner diameter (≈ 100 μm) and hence the absence of peak broadening by lateral diffusion, their resolution is extremely high. It theoretically compares to a distillation column with 100,000 plates [79].

Dimethyl siloxanes are mostly used as stationary phases. They are nonpolar and can be heated to relatively high temperature (350 °C) [80]. Due to the lack of functional groups, analytes are separated mainly based on their boiling point. For more difficult tasks, like the separation of isomers, the polarity of the columns is increased by substitution of methyl by phenyl, cyanopropyl or other more polar groups.

After passing through the column, compounds can be detected by different means. The TCD (thermal conductivity detector) exploits the different thermal conductivities of analyte and carrier. In the FID (flame ionization detector) the carrier gas is mixed with hydrogen and ignited. An electrode measures ions formed upon combustion of carbon containing compounds. Also widespread are MSDs (mass selective detectors), which was the type of detector used in this study.

3.2.2 Mass Spectroscopy

Mass spectroscopy is a powerful tool. It not only detects compounds but also provides information about their structure. The basic operating principle of a mass spectrometer is to ionize molecules and determine the m/z-ratio (mass/charge) of the resulting ions. Two major ionization methods are distinguished [81]. In the EI (electron impact) mode the vaporized molecules are bombarded with a beam of energetic electrons (70 eV). Upon collision, one electron is expelled from the sample molecule, resulting in a molecular radical cation:
$$\mathrm{M} + \mathrm{e}^- \longrightarrow \mathrm{M}^{+\cdot} + 2\,\mathrm{e}^-$$
Since the energy of the electron exceeds the ionization energy of the sample molecule by far, $\mathrm{M}^{+\cdot}$ usually fragmentates, which allows for identification of structural units. These fragmentation patterns are characteristic for each molecule and collected in databases, therefore allowing for the identification of unknown compounds.

The CI (chemical ionization) mode is less energetic. The ionization chamber is equipped with a second inlet for gases like methane or ammonia, which are first ionized and then transfer their charge according to:
$$\mathrm{M} + \mathrm{CH}_5^+ \longrightarrow \mathrm{M}\text{--}\mathrm{H}^+ + \mathrm{CH}_4$$
Therefore, molecules do not fragmentate and their molecular mass can be determined.

For the separation of ions by their m/z-ratio, quadrupole instruments are most widely used. They consist of four parallel metal rods. To these is applied a superposition of a DC (direct current) and RF (radio frequency) potential. For a certain setting only ions of a defined m/z-ratio can pass the rods and reach the detector, whereas the oscillation of the other ions is unstable so that those are deflected towards the rods and discharged. By changing DC and RF settings, a mass spectrum can be scanned. The resolution of a quadrupole is only moderate but it allows the recording of several complete scans in one second. TOF (time of flight) instruments and FT (Fourier transform) spectrometer achieve higher resolution, but due to their costly design they are scarcely used.

One advantage of a quadrupole is that it can be set to monitor only certain ions. This is useful for enhancing sensitivity and therefore recommended for quantification. Furthermore, interfering compounds can be shielded. This mode of operation is called SIM (selective ion monitoring).

Especially when several sample preparation steps, like extraction or derivatization, have to be performed prior to analysis, it is important to take into account a possible loss of sample. Therefore, an *internal standard* (IS) that is chemically similar to the analyte, is added before such steps are carried out. As the peak area is proportional to the analyte concentration in a certain range, its concentration can be calculated based on the tracer, according to

$$KF_i = \frac{f_{\mathrm{IS}}}{f_i} = \frac{c_i \cdot a_{\mathrm{IS}}}{c_{\mathrm{IS}} \cdot a_i} \qquad (3.3)$$

with KF_i being the calibration factor of species i, f_i its response factor, c_i its concentration and a_i the integrated signal area. KF_i can be determined using a calibration curve.

3.3 Analytics of Selected Species

3.3.1 Identification of Unknowns by Silylation

Gas chromatography is restricted to thermally stable and volatile compounds. Unfortunately this excludes most biomolecules. However, these compounds can be converted to suitable derivatives. Therefore, acidic hydrogens (e.g. from hydroxy, carboxyl, thiol and amino groups) have to be replaced by non-polar moieties. Most common methods employ acylation, alkylation or silylation. The latter is extremely useful, as all acidic hydrogens can be exchanged in one step. Even when derivatization is not absolutely necessary, it is often employed to improve peak shape and sensitivity. When the mass spectrometer is operated in SIM mode, characteristic ions of the derivatives can be monitored. Trimethylsilylated compounds for instance always yield an $m/z = 73$ ion $(\mathrm{Si(CH_3)_3^+})$, whose signal can be conveniently employed for sensitive quantification.

Silylation was used for qualitative analysis. Most popular is the insertion of a trimethylsilyl moiety. The silyl group makes most molecules highly thermally stable, so that even carbohydrates and other sensitive biomolecules can be separated at high temperature without decomposition occurring. Due to these advantages many mass spectra from silylated molecules are recorded in databases, which allows for the

Chapter 3 Experimental

identification of unknowns.

The drawback of silylation is the moisture sensitivity of the reagents and derivatives. Therefore, water was removed from the sample by freeze drying or heating in an oil bath at 50 °C under a gentle stream of nitrogen. Traces of water could be removed by subsequent azeotropic evaporation with dichloromethane. Afterwards 500 μL pyridine was used as solvent and then 200 μL of bis-N,O-trimethylsilyl trifluoroacetamide (BSTFA) with 1% trimethylchlorosilane (TMCS) as catalyst were added and the mixture heated at 60 °C for about 10 min. Solutions containing amino acids required heating at 70 °C for 20 min using acetonitrile as solvent [82]. Products were separated on an HP-5ms column. 1 μL was injected in split mode (1:25). The oven temperature program was 50 °C (hold 1 min) — 20 °C/min — 300 °C (hold 1 min) with solvent delay set to 3.6 min.

3.3.2 Carbohydrate Analysis

Carbohydrates can be analyzed both by liquid or gas chromatography. Considering the broad spectrum of products in the formose reaction, gas chromatography was the method of choice due to its superior resolution. There are different ways to convert carbohydrates into volatile compounds, e.g. silylation or acylation [83]. However, direct derivatization leads to the formation of many isomers for a single sugar and thus results in several peaks for each compound. In solution the open chain is in equilibrium with cyclic structures. Derivatization usually occurs much faster than the time necessary for structural equilibration. Therefore, the initial composition is retained upon derivatization. The presence of multiple peaks, furthermore appearing in different ratios, greatly hampers quantification.

This requires a prior step, namely the modification of the carbonyl group to break down this equilibrium, e.g. by the formation of oximes [84]. Depending on whether *syn* or *anti* isomer are resolved, either one or two peaks are obtained. However, this method permits the unambiguous identification of carbohydrates without loss of information.

Another method is the reduction of the carbonyl group to convert carbohydrates to their corresponding polyols, which can be conveniently performed with $NaBH_4$ in aqueous solution. The inability of distinguishing aldoses and ketoses does not

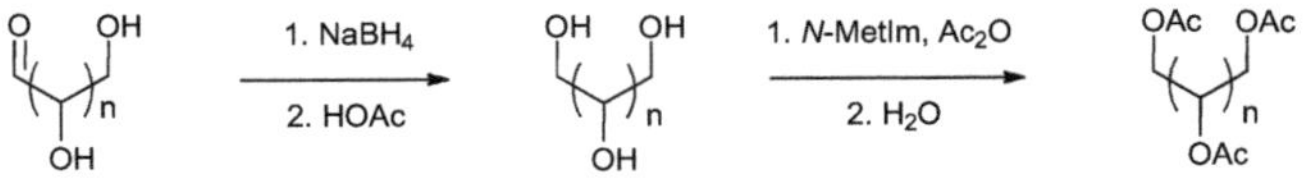

3.3 Analytics of Selected Species

Figure 3.3: Preparation of alditol acetates from carbohydrates.

have to be a drawback, as this reduction step drastically minimizes the number of possible derivatives. There exist only 13 linear alditols (reduced sugars) with two to six carbon atoms (enantiomers not discriminated), whereas there are much more possible precursors. Due to the large number of different products in the formose reaction, this procedure was chosen for the work presented here.

In a second step the alditols were acetylated to obtain volatile compounds. Acetylation was chosen because it tolerates a certain amount of water, which is simply removed by an excess of acetic anhydride. Furthermore, salts that were present in our mixtures did not inhibit the reaction. Silylation in contrast was sensitive to certain salts. Alditol acetates are quite stable and could be stored several weeks in a freezer.

The presence of alditol complexing borate species blocks acetylation with acetic anhydride, so it usually has to be removed first. However, with N-methyl imidazole as catalyst it is possible to perform acetylation in the presence of borate (figure 3.3) [85]. The disadvantage is the occurrence of some side products, originating from the catalyst, which prevents trace analysis by splitless injection. The solution was therefore concentrated prior to the acetylation step, and split injection was chosen to obtain clear chromatograms (see reference chromatogram figure 3.4).

Prior to carbohydrate analysis *myo*-inositol was added as internal standard. Afterwards carbohydrates were reduced to alditols using $NaBH_4$. A sample resulting from the reaction of 50 mg formaldehyde was reduced with 500 µL of a solution of 1 g $NaBH_4$ in 6 mL 1.7 M NH_3 for 2 h at 40 °C. Excess hydride was destroyed by addition of 500 µL acetic acid. Following reconcentration to about 1 mL, 200 µL were taken and acetylated with 2 mL acetic anhydride and 200 µL N-methyl imidazole for 10 min at room temperature. The solution was quenched with 5 mL water and after decomposition of excess acetic anhydride the alditol acetates were extracted with 1 mL dichloromethane, which then was dried over anhydrous Na_2SO_4. This

Chapter 3 Experimental

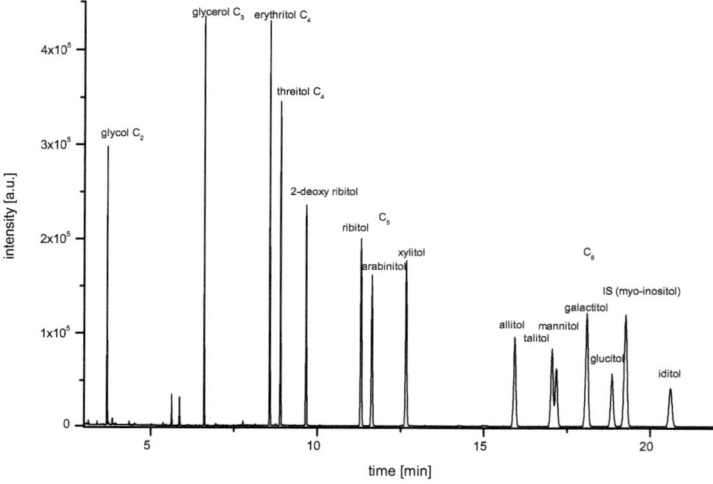

Figure 3.4: Gas chromatogram of an alditol acetate reference mixture showing high resolution even of stereoisomers.

extraction process ensures separation from perturbing compounds, which remain in the aqueous phase.

The ions $m/z = 43$, 115 and 129 were scanned, each with a dwell time of 50 ms. Neglecting other ions reduces interference and enhances sensitivity. The ion $m/z = 43$ is the CH_3CO^+ ion, originating from the acetyl groups, whereas $m/z = 115$ is a characteristic signal of carbohydrates and $m/z = 129$ of deoxysugars. Calibration was performed with reference sugars. Alditols were prepared from the corresponding aldoses, except for erythritol, threitol and iditol, which were directly employed as such. With a highly polar DB-225ms phase (50%-cyanopropyl-phenyl-50%-dimethylsiloxane) it was possible to obtain a high resolution, even of the different stereoisomers, as shown in figure 3.4. All potential linear alditols with two to six carbon atoms were separated sufficiently. 1 μL sample was injected in split mode (1:50). The oven temperature program was 100 °C (hold 1 min) — 20 °C/min — 230 °C (hold 20 min) with solvent delay set to 2.5 min.

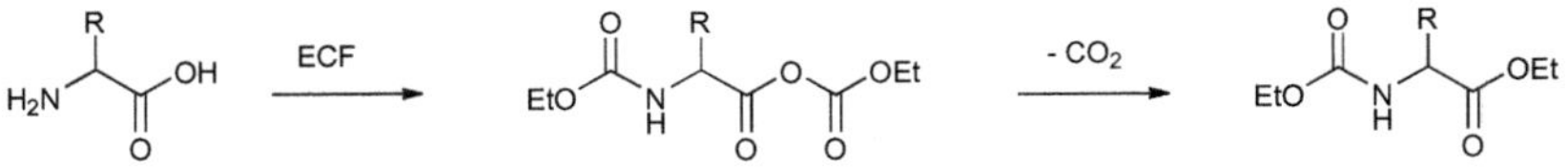

3.3 Analytics of Selected Species

Figure 3.5: Derivatization of amino acids with ethyl chloroformate yields volatile products.

3.3.3 Amino Acid Analysis

The analysis of amino acids also requires additional modification steps. One suitable technique is HPLC in combination with either pre- or post-column derivatization. Afterwards, analytes can be detected by fluorescence or absorption. When performing GC, both functional groups (amino and carboxylic acid) and the polar side-chains must be modified to obtain volatile derivatives. In principle silylation is possible. However, elevated temperatures are necessary to ensure complete reaction. Furthermore, some amino acids give rise to two peaks, hampering quantitative analysis. Glycine for instance can form di- and tri-TMS derivatives [86].

Another method was therefore developed. Amino acids were converted to volatile products by derivatizing amino and carboxylic acid groups in a one step reaction with ethyl chloroformate (ECF) [87]. This protocol is superior as it even works in aqueous phase, which allows for purification from complex matrixes, as the resulting derivatives can be extracted into an organic layer. This makes it especially useful for analysis of biological samples. Compared to silylation sensitivity, reliability and economy of time are improved. The only drawback concerns histidine and arginine. Their side chains are not functionalized, so that these two amino acids cannot be detected by the ECF method.

For derivatization 400 μL of a 4:1 mixture of ethanol:pyridine were added to 600 μL of aqueous sample. Valine was used as internal standard. 50 μL ethyl chloroformate were then added and the vial briefly shaken, during which amino acids are converted to N-ethoxycarbonyl ethyl esters. The intermediate ethoxycarbonyl esters decarboxylate instantaneously, catalyzed by pyridine (figure 3.5). After sonication for a few seconds, the resulting derivatives were extracted with 1 mL chloroform containing 1% ethyl chloroformate. 1 μL was injected in split mode (1:25). Products were separated on an HP-5ms column. The oven temperature program was 50 °C (hold 1 min) — 20 °C/min — 300 °C (hold 1 min) with solvent delay set to 3.6 min.

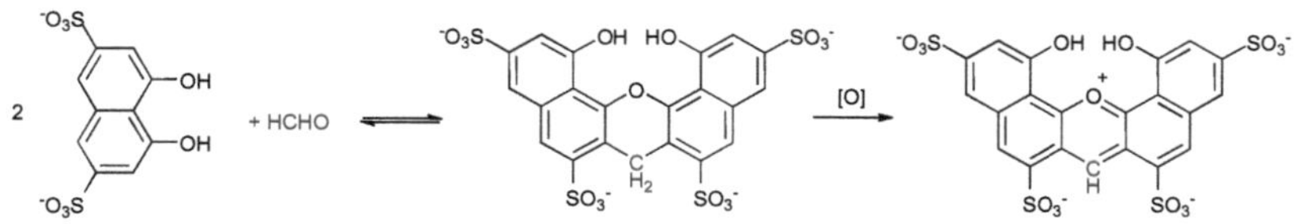

Figure 3.6: Proposed mechanism for the chromotropic acid assay in concentrated sulphuric acid.

3.3.4 Formaldehyde Analysis

Formaldehyde, used as precursor to synthesize carbohydrates, can in principle also be measured by gas chromatography. However, due to the presence of salt, solution could not be injected directly. Therefore, the conversion of formaldehyde was monitored photometrically.

The assay uses chromotropic acid, which forms an adduct with formaldehyde that subsequently is oxidized by concentrated sulphuric acid to a purple dye. The exact mechanism and product are still controversial [88, 89]. Most plausible is a pathway as shown in figure 3.6. The first step is an addition of formaldehyde to the chromotropic acid, whereby this adduct can link to another molecule. This condensation is similar to the formation of phenol formaldehyde resins. Sulphuric acid acts as water scavenger and likewise as oxidant for the second step. The formed cation is energetically favoured due to the large aromatic system. Different products have been claimed in literature as well. The chromotropic acid assay is characterized by high sensitivity and selectivity. However, there were slight interferences with formose products, which was probably the reason why absorbance did not reach the baseline even for prolonged reaction times.

The chromotropic acid reagent was prepared by dissolving 0.5 g of chromotropic acid disodium salt dihydrate in 3.75 mL of water and subsequent addition of 100 mL concentrated sulphuric acid [90]. To determine the concentration of formaldehyde the sample was diluted with water to a maximum of 3 mM HCHO. 100 μL were reacted with 1 mL of reagent solution in an oil bath at 100 °C for 15 minutes. After cooling to room temperature and dilution with 5 mL water the absorbance was read off at 578 nm against a blank solution, prepared by the same protocol with pure water. The concentration of formaldehyde was calculated using a calibration curve.

Chapter 4

Prebiotic Carbohydrate Synthesis

4.1 Background

As nearly 3.5 billion year old microfossils and typical isotope patterns in sediments indicate, life arose quite early on earth [91]. Astonishingly, the primary synthesis of biomolecules obviously took place at high speed and environmental conditions very different to the current ones. Temperatures in the ocean in this hadean period were around 200 °C at a pressure of 20 bars or higher [92], and a reducing atmosphere together with a flux of energy in the form of electric discharge (lightnings), radiation and volcanic activity could induce the formation of organic molecules from simple gases like CH_4, NH_3 or H_2 and water. This was first shown by Stanley Miller in 1953 [93, 94]. He observed the formation of amino acids, which led to the speculation of whether proteins were the basis for the origin of life [95]. The list of potentially available molecules has been expanded ever since [96]. Later, strengthened by the discovery of catalytically active RNA, the RNA world hypothesis was postulated. Nucleic acids can also be synthesized from a possibly prebiotic atmosphere [97, 98].

Another important class of biomolecules, the carbohydrates, were equally formed in such experiments. They arise from the condensation of formaldehyde, which itself can be synthesized by radiation or electric discharge in considerable yield. Therefore, it is generally viewed as a prebiotic molecule [99]. The finding of formaldehyde condensation dates back to 1861. Butlerow observed the formation of a sugar-like compound when formaldehyde reacted in $Ca(OH)_2$-solution [100]. This condensation is called the *formose* reaction. Due to the ease with which complex carbohydrates are

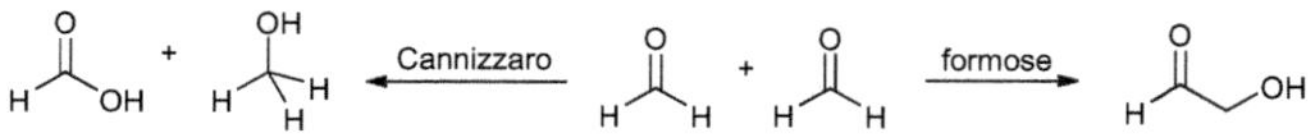

Chapter 4 Prebiotic Carbohydrate Synthesis

Figure 4.1: In basic media the disproportionation of formaldehyde competes with C-C bond formation.

synthesized from a very simple precursor, the formose reaction has been considered to have contributed to the origin of life. However, due to the fast degradation and the missing selectivity, this is very doubtful [101–103].

First studies in the formose reaction were, however, not motivated by its prebiotic relevance. Carbohydrates are a major nutrient and the formose reaction is an easy means for their production on an industrial scale. In contrast to other synthetic pathways just one step is required. However, the variety of products prevents nutritional purpose and separation is laborious [104]. Nowadays, the idea to produce carbohydrates from fossil resources, as it was proposed, may seem strange to us. In those days however, green chemistry was not a big issue, and oil was cheap. Today, such a synthesis would of course be both economically and environmentally unviable.

The kinetics of the formose reaction is quite complex. In basic media, where it occurs, formaldehyde does not easily react with itself to establish a carbon-carbon bond. Rather the Cannizzaro reaction takes place (see figure 4.1). However, some condensation product glycoladehyde (GA) is formed as well, which initiates a cascade of reactions, ultimately leading to the formation of various straight-chain and branched carbohydrates. GA can be deprotonated by bases and thus an nucleophilic carbon center is created. This enediol is stabilized by certain cations, e.g. Ca^{2+} to form an intermediate, which then reacts with formaldehyde in an aldol reaction (see figure 4.2). By subsequent isomerizations (also via the enediol complexes) and condensations larger carbohydrates are built up. In addition, retro aldol cleavages may

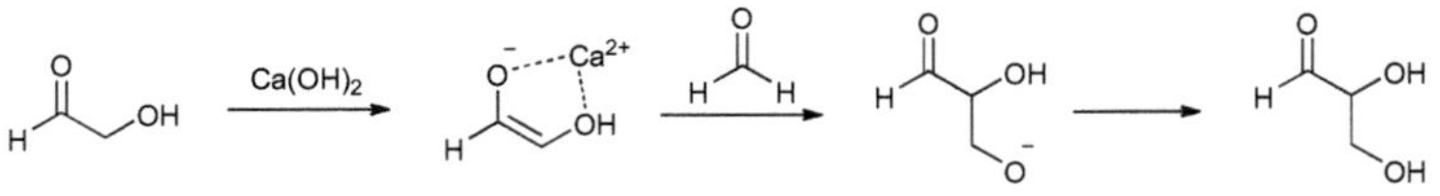

Figure 4.2: $Ca(OH)_2$ acts as base and catalyst by stabilizing the enediol form of carbohydrates, which then condensate with formaldehyde.

4.1 Background

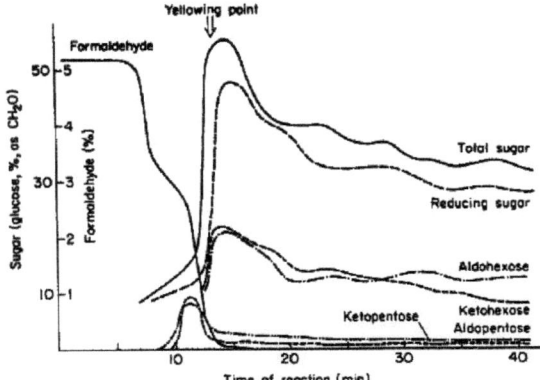

Figure 4.3: Sugar formation in a typical formose reaction (2 M formaldehyde in 0.2 M Ca(OH)$_2$ at 60 °C), taken from [104].

occur, which lead to the formation of two smaller carbohydrates from a bigger one. These two fragments can reinitiate the reaction, so that the formose reaction evolves to an autocatalytic system.

At the beginning, when no GA is present, only the Cannizzaro reaction takes place. Formaldehyde is consumed slowly in an induction period, but as soon as some GA formed, the formose reaction is initiated and the conversion of formaldehyde starts to accelerate. The main products are pentoses and hexoses. They accumulate by forming stable rings, which slows down retro aldol cleavages. However, carbohydrates are not stable in basic media but decompose to polymeric humins. This side reaction causes a yellowing of the solution, which occurs shortly before reaching maximum yield. The formose reaction is hence characterized by a lag period, the formation and finally the decomposition of carbohydrates. The lag period can be shortened by adding small sugars like GA as initiators, though they will not increase the yield of carbohydrates, but solely shorten reaction times. Due to the formation of humins, the yield usually does not exceed 50%. A typical kinetics is shown in figure 4.3.

Since the formose reaction is an aldol condensation, it occurs at high pH. Performing the reaction in NaOH will not yield any carbohydrates. Additionally, catalytically

Chapter 4 Prebiotic Carbohydrate Synthesis

active cations are necessary. They form complexes with the enediol form of carbohydrates and thus facilitate condensation reactions [105]. Often $Ca(OH)_2$ is used as both catalyst and base due to its high activity. Apart from this base, many other catalysts have been identified [104]. Naturally occurring minerals and clays can also evoke the formation of carbohydrates when refluxing a formaldehyde solution [106–108].

Contrary to what might be a conventional property of catalysts, the use of different bases in the condensation of formaldehyde hardly influences selectivity. The outcome is always a broad spectrum of different sugars. The addition of initiators does not improve selectivity either. This concern and the problem of low yield can be addressed by employing other small molecules instead of formaldehyde to build up carbohydrates. Using short sugars, like glycolaldehyde or glyceraldehyde, other catalysts and less harsh conditions are sufficient. Examples include zinc prolate [109], silicate [110] or dipeptides [111]. With the latter even an enantiomeric excess could be achieved. However, in none of these cases a successful formose reaction, using solely formaldehyde, could be demonstrated.

In our approach, experiments were restricted to simple hydrothermal conditions. Using formaldehyde without additional initiators in the presence of only simple salts, reaction sequences were analyzed. The formose reaction is usually performed at moderate temperatures and occasionally at around 100 °C [112]. The motivation for studying the formose reaction in high-temperature water is based on the lack of data on the hydrothermal behaviour of formaldehyde, but also on the fact that early terrestrial conditions might have included various similar aqueous environments.

4.2 Moderate-Temperature Reaction

For comparison, the formose reaction was first performed at moderate temperatures. As the typical product spectrum is rather broad, the analysis of carbohydrates is hampered. In our approach all linear carbohydrates could be quantified by gas chromatography. This technique offers superior separation and simultaneous removal of high molecular weight decomposition compounds, which could interfere with analysis. In literature different means of identification and quantification have also been

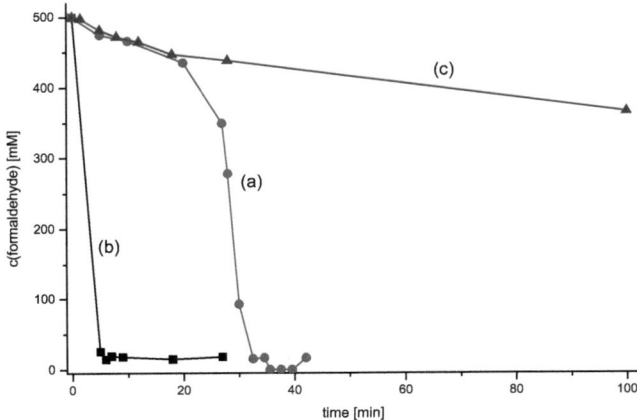

Figure 4.4: Conversion of formaldehyde at 60 °C in a) 0.05 M Ca(OH)$_2$; b) 0.05 M Ca(OH)$_2$ with 1 mol% GA and c) 0.1 M NaOH with 1 mol% GA.

employed, which are less reliable, e.g. paper chromatography or photometric assays. Furthermore, the yield depends on the reagent concentration. In order to compare the hydrothermal reaction with moderate temperature, experiments were carried out at 60 °C. A 0.5 M formaldehyde solution was reacted in the presence of hydroxides. The same concentration of formaldehyde was used for hydrothermal experimentation. As catalyst 0.05 M Ca(OH)$_2$ or 0.1 M NaOH was added, so the concentration of OH$^-$ was equal in both cases.

The conversion of formaldehyde under Ca^{2+}-catalysis is shown in figure 4.4(a). Initially the conversion was slow. After it started to accelerate, a process initiated by the formation of glycolaldehyde, the initial turbid solution (concentration of Ca(OH)$_2$ is higher than its solubility) turned clear. This can be attributed to the formation of soluble enediol complexes. As soon as some condensation product formed, the reaction ran quickly until the consumption of formaldehyde was complete, and the solution turned yellow. Figure 4.4(b) was obtained when 1 mol% GA was initially present in addition. Here the conversion started immediately. The same physical observations could be made, i.e. the solution became clear and then turned yellow.

Chapter 4 Prebiotic Carbohydrate Synthesis

With substitution of $Ca(OH)_2$ by NaOH as base (figure 4.4(c)) there was only a slow formaldehyde consumption, even though an initiator was added. The rate of conversion is comparable to that during the induction period in the $Ca(OH)_2$-catalyzed mixture. In both cases only the Cannizzaro reaction seems to occur. The addition of initiator obviously has no impact when NaOH is present. GC analysis of this reaction revealed not only the absence of carbohydrates, but also the decomposition of the added glycolaldehyde. Consequently, the presence of catalytically active ions is crucial at moderate temperatures.

Analysis of the reaction mixtures in the presence of $Ca(OH)_2$ verified the formation of carbohydrates. Sugars were quantified as their alditol acetates, whereby the number of different compounds was minimized by an additional reduction step with $NaBH_4$, thereby simplifying analysis. Aldoses yield the same product as ketoses with corresponding stereochemistry. Only the linear carbohydrates were quantified. It is known that branched sugars are also formed in formose reactions and the gas chromatograms show some peaks that could be attributed to them. The total yield is therefore underestimated in our experiments, as those are not taken into account.

The kinetics in absence of initiator is shown in figure 4.5. For clarity, isomers are merged. During the induction period no sugars were present. The formation of carbohydrates started abruptly after 25 min. The time required to reach maximum yield was much shorter than the induction period itself. Furthermore, the different sugars were not synthesized uniformly. When comparing the maxima one notices that the smaller carbohydrates appear prior to the larger ones. This reflects a gradual assembly of the carbohydrates by successive additions of formaldehyde. The total sugar yield passed through a maximum. 26.5% of linear carbohydrates were obtained at best. The composition at the maximum is shown in table 4.1. Major products are hexoses and pentoses, which accumulate because of their relative stability. Probably due to a difference in energy, stereoisomers are not formed in equal amounts.

Performing the reaction in presence of 1 mol% GA did not improve yield. Also, the relative product distribution was unchanged. This confirms that although an initiator drastically reduces the lag period, it is incapable of governing selectivity and improving yield. Keeping these results in mind, we now focus on the formose reaction in HTW.

4.3 Hydrothermal Formose Reaction

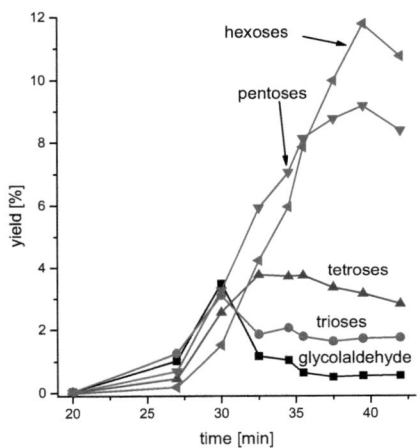

sugar	alditol	yield [%]
GA	glycol	0.55
trioses	glycerol	1.73
tetroses	erythritol	1.54
	threitol	1.62
pentoses	ribitol	1.65
	arabinitol	4.27
	xylitol	3.25
hexoses	allitol	0.88
	talitol	2.01
	mannitol	1.37
	galactitol	0.86
	glucitol	4.38
	iditol	2.29
total		26.5

Figure 4.5: Kinetics of sugar formation at 60 °C, catalysis by Ca(OH)$_2$.

Table 4.1: Carbohydrate composition at reaction time with maximum yield

4.3 Hydrothermal Formose Reaction

4.3.1 Effect of Added Salt

For hydrothermal processing, the formaldehyde solution was simply heated at 200 °C. The conversion of formaldehyde in different salt solutions is shown in figure 4.6. In pure water and even under acidic conditions (such as diluted acetic acid), formaldehyde is consumed relatively fast within a timescale of minutes. An induction period, as described for the formose reaction at moderate temperatures, is not observed. With increasing basicity of the added salts, the conversion is accelerated. Even the barely basic sodium sulfate shows some effect. This trend is continued following the series acetate, hydrogen carbonate and hydrogen phosphate. In a carbonate buffer (50 mM NaHCO$_3$, 50 mM Na$_2$CO$_3$), formaldehyde is consumed in less than one minute.

Of course, the fact that formaldehyde vanishes does not prove the formation of

Chapter 4 Prebiotic Carbohydrate Synthesis

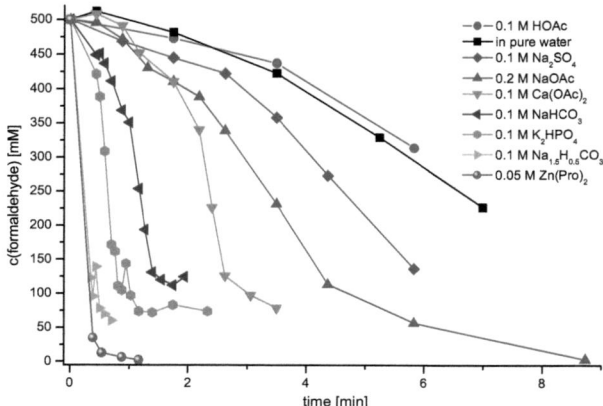

Figure 4.6: Conversion of formaldehyde at 200 °C and 100 bar in the presence of various salts.

carbohydrates. In fact, the NMR spectra of the reactions performed in pure water, in acetic acid and also when sodium acetate was added, just show the Cannizzaro products formic acid and methanol (figure 4.7). These solutions remained clear and colourless even for prolonged reaction times and did not show the typical yellowing point, resulting from decomposition products of carbohydrates. This colour change was, however, observed for all more basic salts, indicating a successful formose reaction despite the absence of catalytically active cations.

4.3.2 Product Identification

To gain insight into the composition of formose products, different analytical techniques were applied. Besides formaldehyde the Cannizzaro products methanol and formic acid are also volatile. Nonvolatile compounds include carbohydrates but also their decomposition products. Polymeric species as well as acids, which could be formed via Crossed Cannizzaro reactions, can be expected.

To evaluate the different fractions, formaldehyde was reacted in 0.1 M $NaHCO_3$ in the presence of 1 mol% dihydroxyacetone as initiator for 0.5 min. This is at the yellowing point, where the highest yield of sugars can be expected. The con-

4.3 Hydrothermal Formose Reaction

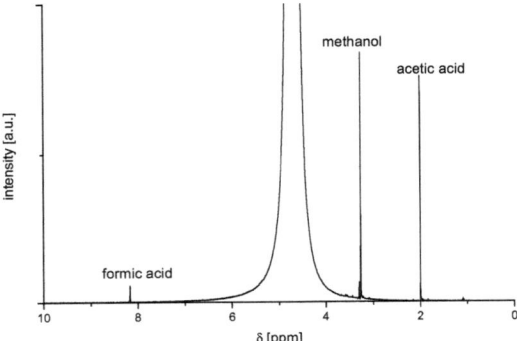

Figure 4.7: NMR spectrum of the hydrothermal formose reaction in 0.1 M acetic acid just shows Cannizzaro products; reaction time 5.8 min at 200 °C.

version of formaldehyde was 70%. The fraction of nonvolatile compounds was determined by weighing after freeze drying. They accounted for 61% with respect to initial formaldehyde and dihydroxyacetone. Thus the remaining 9% must be other volatile compounds. This simple analysis by weighing must be interpreted with care. Formaldehyde can polymerize to paraformaldehyde or cyclic oligomers, which are no longer volatile. Besides, acids can be fixed in the nonvolatile fraction as their salts. A less controversial analysis by pH measurements will be presented later.

For further analysis the nonvolatile fraction was resolubilized in D_2O. Its ^{1}H-NMR spectrum is shown in figure 4.8. For better tracing of decomposition products the reaction time was extended to 2 min as comparison. Here the solution was dark brown and exhibited a strong caramel-like odour. The sample after 0.5 min shows many signals between 3.5–4.5 ppm. These are typical for carbohydrates and correspond to protons attached to a carbon with a hydroxy group. Also, some small signals are present at around 5 ppm. They may originate from protons at the anomeric centers. Due to their ability to form cycles, carbohydrates do not show signals at low field, where aldehyde protons usually appear. The peak at 8.4 ppm is probably attributed to formate, which is fixed as its sodium salt in the nonvolatile fraction. The absence of peaks in the range between 6–8 ppm indicates that no aromatic compounds or

Chapter 4 Prebiotic Carbohydrate Synthesis

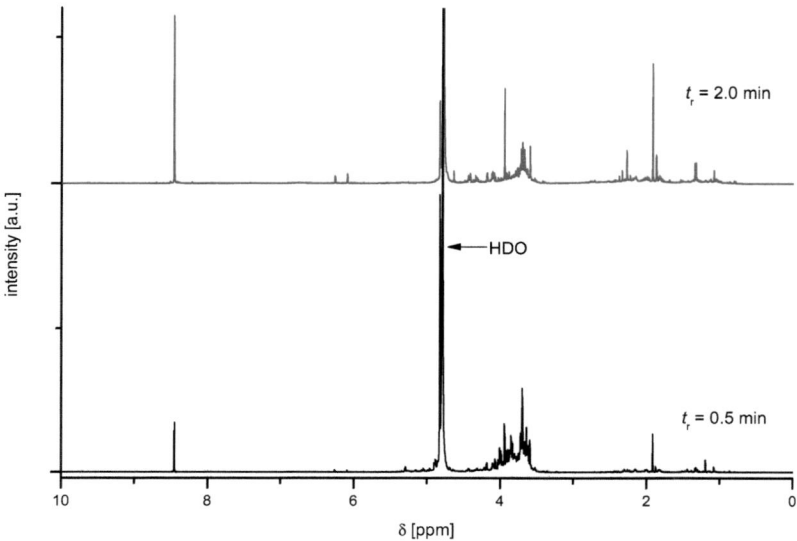

Figure 4.8: ^{1}H-NMR spectra of a formose reaction in 0.1 M NaHCO$_3$ at 200 °C with 1 mol% dihydroxyacetone added; the lower spectrum was obtained after 0.5 min, around the yellowing point and the upper one after 2 min; spectra recorded in D$_2$O.

furans were formed. The presence of peaks at low chemical shift between 1–2 ppm is striking as they cannot be expected from simple carbohydrates. These are typical alkane protons indicating the presence of deoxy compounds.

At longer reaction times two small signals appear at around 6 ppm in the olefinic region. The signals at low chemical shift are more pronounced. Whereas the ratio of the integrals from the signals at around 4 ppm relative to the 1–2 ppm region is 10:2 at the yellowing point, it is 10:9 at prolonged reaction times. This indicates strong chemical changes upon decomposition.

Analogous conclusions can be drawn from the ^{13}C-NMR spectra shown in figure 4.9. Carbohydrate signals are usually between 60–100 ppm. In the hydrothermal formose reaction many peaks also lie in this range. Again, no aromatics and furans can be detected, which would show signals between 100–150 ppm. Some peaks appear

4.3 Hydrothermal Formose Reaction

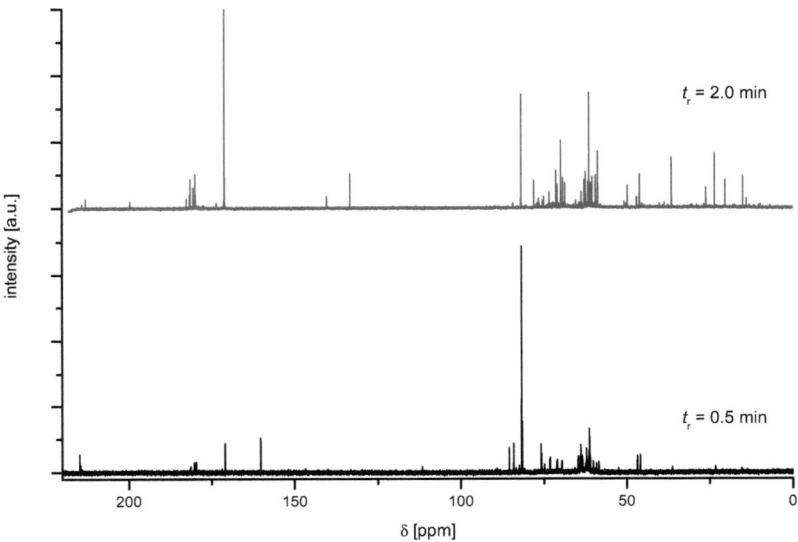

Figure 4.9: ^{13}C-NMR spectra of a formose reaction in 0.1 M NaHCO$_3$ at 200 °C with 1 mol% dihydroxyacetone added; the lower spectrum was obtained after 0.5 min, around the yellowing point and the upper one after 2 min; spectra recorded in D$_2$O.

at higher chemical shift, probably carbonyl as well as carboxylic acid groups. In consistency with the ^{1}H-NMR spectrum some peaks are present at low chemical shift below 50 ppm, which again are due to saturated hydrocarbons. At longer reaction times more peaks appear in this region.

Due to the large number of signals, no attempts were made to assign them to certain compounds. Furthermore, the presence of peaks in the typical carbohydrate region does not necessarily prove that solely sugars were formed. Decomposition products probably show similar structural units with equivalent chemical shifts. Hence, gas chromatographic measurements were the only means to unambiguously verify the presence of carbohydrates.

As we first reduced the carbohydrates with NaBH$_4$, one could argue that sugar alcohols might have already been present in the initial mixture. To prove the si-

Chapter 4 Prebiotic Carbohydrate Synthesis

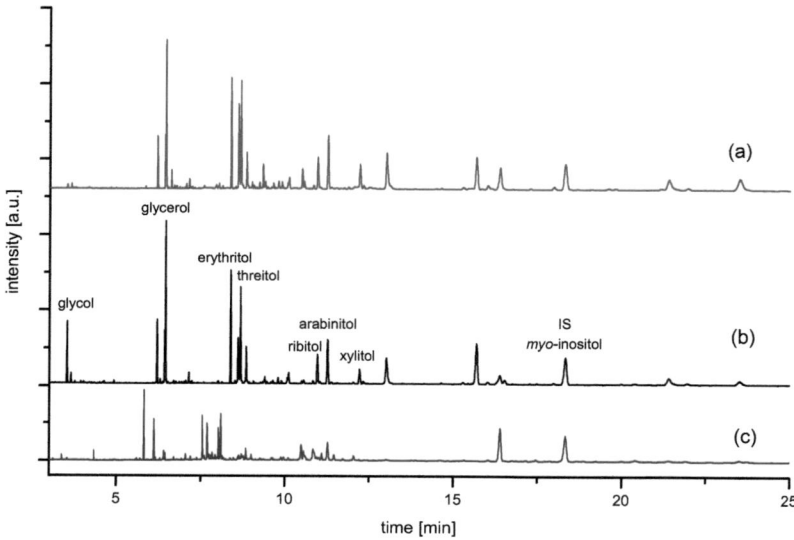

Figure 4.10: GC traces of a 0.5 M formaldehyde solution in 0.1 M K_2HPO_4 reacted for 0.81 min at 200 °C and 100 bar; a) after reduction with $NaBH_4$, esterification with methanolic HCl and acetylation; b) after reduction with $NaBH_4$ and acetylation; c) directly acetylated reaction mixture. Only marginal amounts of sugar alcohols are present after reaction.

multaneous presence of carbonyl as well as hydroxy groups in the products, one characteristics of carbohydrates, gas chromatographic measurements were performed after different derivatization procedures. A formose reaction at its yellowing point was investigated. A chromatogram obtained after the typical derivatization protocol (reduction with $NaBH_4$ and subsequent acetylation) is shown in figure 4.10(b). The peaks of the linear alditols are assigned. Besides them also other species can be detected. When the reduction step was omitted and the reaction mixture directly acetylated, the chromatogram 4.10(c) was obtained. Peaks that are present in both chromatograms ((b) and (c)) must be compounds that do not contain any carbonyl group. Otherwise their position would be shifted as carbonyl containing species are modified by $NaBH_4$ reduction. Indeed, both chromatograms are very different. Only

4.3 Hydrothermal Formose Reaction

marginal amounts of sugar alcohols were present in the initial mixture. This proves that Crossed Cannizzaro reactions do not occur to a significant extent. Furthermore, peaks in the reduced sample seem to be shifted towards longer retention times. This is reasonable, as reduction and acetylation of the newly created hydroxy group increases molecular weight and thus retention time. This allows the conclusion that the predominant fraction of the alditols indeed originate from the reduction of the corresponding carbohydrates.

Apart from carbonyl and hydroxy groups, one might also expect the formation of carboxylic acids. Due to their high polarity they are poorly resolved by gas chromatography. Therefore, a sample first reduced with $NaBH_4$ was heated three times with methanolic HCl, which was evaporated after each cycle. Thereby carboxylic acids are converted into methyl esters, which are easily separated by GC. The chromatogram after acetylation is shown in figure 4.10(a). It largely corresponds to that of the sample which was not esterificated. Only few additional peaks appear. This leads to the overall conclusion that the compounds we can observe by GC contain hydroxy groups and (at least) one carbonyl but no carboxylic acid groups. In summary, the common technique of a prior reduction step is useful for minimizing the number of species without producing artifacts.

4.3.3 Influence of Catalytically Active Species

We also tested for the influence of both the cation variation and the presence of an initiator. A 0.2 M sodium acetate or a 0.1 M calcium acetate solution, respectively, were used to investigate the effect of a potential cation catalysis. Both solutions contain the same amount of acetate and thus exhibit similar basicity. Figure 4.11 shows that up to two minutes reaction time the consumption of formaldehyde is identical in both cases. Subsequently, the reaction accelerates in the calcium acetate solution. An interpretation could be that at this point small carbohydrates such as glycolaldehyde must have formed, which can specifically interact with the Ca^{2+}-ions and induce the formose reaction. Testing the reaction solutions from the first two minutes, indeed predominantly the Cannizzaro products were found in both solutions. In case of the calcium salt, the solution turns yellow at later stages, and carbohydrates are formed. The maximum yield of carbohydrates is already reached

Chapter 4 Prebiotic Carbohydrate Synthesis

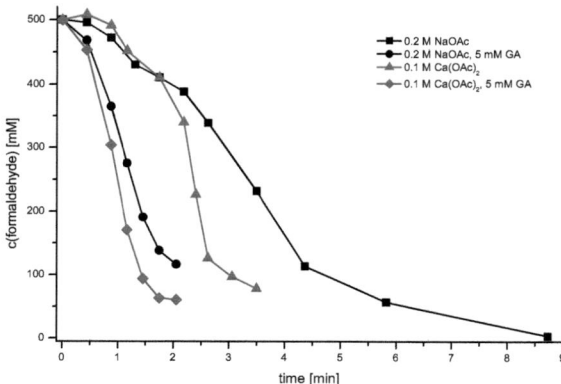

Figure 4.11: Conversion of formaldehyde with addition of sodium (black curves) or calcium (gray curves) acetate in presence and absence of the initiator glycolaldehyde.

for a formaldehyde conversion of 55%. This has to be compared with the formose reaction at moderate temperatures, where the yield is at a maximum shortly after all formaldehyde is consumed. According to the data, decomposition of carbohydrates seems to be even more accelerated at higher temperatures, which is expected. The overall yield of sugars is rather small. Integrating over the signals of all linear sugars with two to six carbon atoms, we obtain only 3.4%. The composition and distribution of sugars will be discussed later.

To compare this salt catalysis, reactions were carried out with the addition of glycolaldehyde as a promoter (1 mol% with respect to formaldehyde). Indeed, the conversion of formaldehyde is much faster (figure 4.11). Here, the sodium acetate solution also turns yellow, and carbohydrates are formed. In case of Na(CH$_3$COO) the total yield is 5.1%, whereas for Ca(CH$_3$COO)$_2$ it reaches 12.0%. Again the maximum of carbohydrate formation is reached well before complete formaldehyde consumption.

At moderate temperatures the presence of Ca^{2+} is crucial for a successful formose reaction. Even with initiator, no carbohydrates are formed in the absence of this catalyst. Under hydrothermal conditions, however, initiator and catalyzing counterion simply step into the competition of formose and Cannizzaro reaction. In the presence of glycolaldehyde the carbohydrate formation starts immediately, whereas in the absence a substantial amount formaldehyde is lost before by self disproportionation. To keep the experiments limited to simple chemicals, the use of an initiator was omitted in further studies. The more basic salts $NaHCO_3$ and K_2HPO_4 or a sodium carbonate buffer also resulted in a yellowing of the solution after relatively short reaction times (figure 4.6) and coupled carbohydrate formation. Not only the speed of formaldehyde consumption, but also the yield of carbohydrates was improved compared to the addition of more neutral salts, such as $Ca(CH_3COO)_2$. This again points to the fact that a slightly more basic pH is of greater relevance than the specific presence of Ca^{2+} under hydrothermal conditions.

4.3.4 Characteristics of the Hydrothermal Reaction

For the model case of K_2HPO_4 addition, the kinetics of formation and decomposition of carbohydrates were analyzed in detail. Figure 4.12 shows the concentration of formaldehyde and the products as added masses of carbohydrates with the same number of carbon atoms. The induction period is only around 25 s. During this time span about 20% of formaldehyde is lost via disproportionation. Afterwards the formation of carbohydrates is very rapid. It takes less than another 30 s to reach maximum yield.

Contrary to the reaction at ambient temperatures, where hexoses are the main product, here the shorter carbohydrates are preferentially formed. Despite the high stability that is in principle expected for hexoses, their amount is insignificant at 200 °C. With the exception of glycolaldehyde, all different carbohydrates peak around the same reaction time, contrary to the control reaction at 60 °C. We speculate that under hydrothermal conditions we have a very fast equilibration between all the different compounds. Using glycolaldehyde or dihydroxyacetone as initiator did not alter product distribution, as compared to the reaction without initiator. The ineffectiveness of different initiators to influence selectivity is well described for the formose reaction.

Not only the formation, but also the decomposition is very rapid. One intermediate seems to be glycolaldehyde itself, because its concentration still increases while all other sugars diminish. The nature of the decomposition products was however not

Chapter 4 Prebiotic Carbohydrate Synthesis

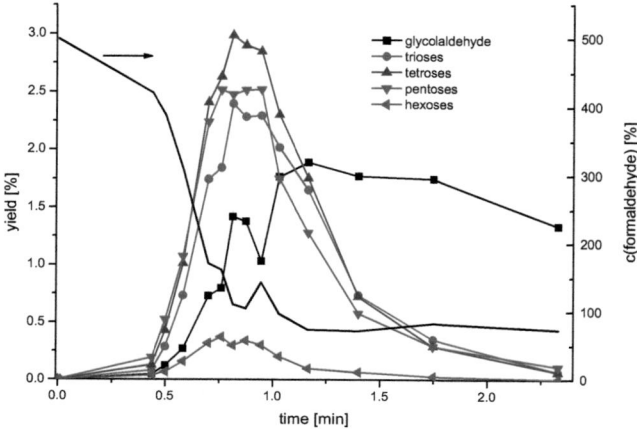

Figure 4.12: Kinetics of formaldehyde consumption and product formation of a 0.5 M formaldehyde solution in 0.1 M K_2HPO_4 at 200 °C.

investigated further. It is known that some polymeric species are formed. At long reaction times the solution turned dark brown and ultimately turbid and exhibited a strong caramel like odour.

Gas chromatographic analysis revealed the presence of other compounds besides straight-chain carbohydrates. The chromatogram of the reaction at maximum yield is shown in figure 4.13. Three different ions were scanned. The signal at $m/z = 43$ originates from the acetoxy cation. Thus all compounds with hydroxyl groups show a peak in this trace. The ion $m/z = 115$ is characteristic for alditols. It originates from a fragment with 3 carbon atoms after loss of acetic acid and ketene [113]. The ion $m/z = 129$ is formed the same way when an additional methyl group is present. Consequently, deoxysugars show this signal. Also in this GC trace some extra peaks are present. Most importantly, we were able to quantify 2-deoxyribitol. Its identity was validated by comparing the complete mass spectrum and the retention time with the reference compound. The maximum yield was however only 0.18 mg per 100 mg formaldehyde. Still, it is worth noting that not only the true condensation products of formaldehyde are found. As the various peaks with the $m/z = 129$

4.3 Hydrothermal Formose Reaction

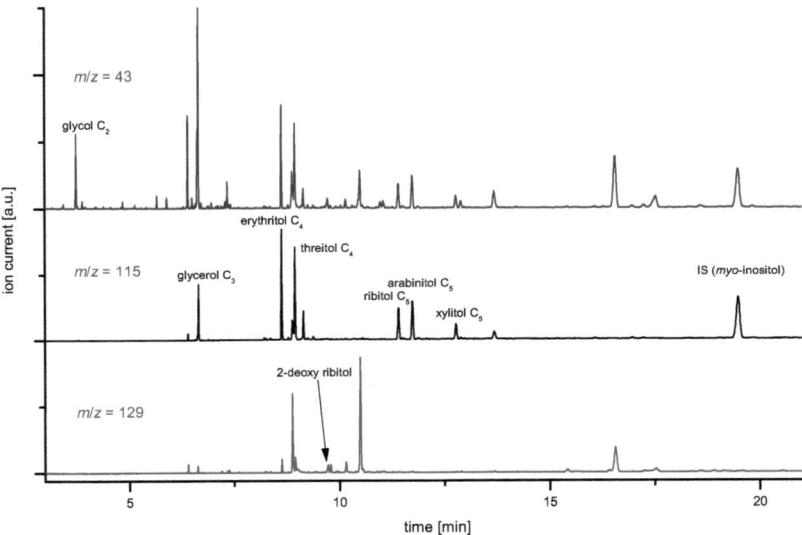

Figure 4.13: GC traces of the reduced and acetylated formose products for different ions (each scaled with different factor).

ion fragment indicate, many other deoxy compounds are probably formed as well. Further indications for the presence of deoxy species are gained from NMR spectra that show peaks in the region of 1–2 ppm (see figure 4.8, page 58). In the control experiment at 60 °C in $Ca(OH)_2$ we also detected 2-deoxyribitol after reduction with $NaBH_4$. Here, the maximum yield was lower (0.11 mg per 100 mg formaldehyde). The decomposition of 2-deoxyribitol is slower than that of the other carbohydrates. This is a general observation for all compounds showing a fragment with $m/z = 129$.

To evaluate the amount of self disproportionation of formaldehyde during the reaction, pH measurements were conducted after the synthesis for the model reaction in 0.1 M K_2HPO_4. The variation of pH (after threefold dilution) is shown in figure 4.14. As expected the pH decreases throughout the whole reaction process. Although no sugars were formed in the first 30 s, the pH falls rapidly. Under the assumption that the change in pH is only caused by the formation of formic acid, its concen-

Chapter 4 Prebiotic Carbohydrate Synthesis

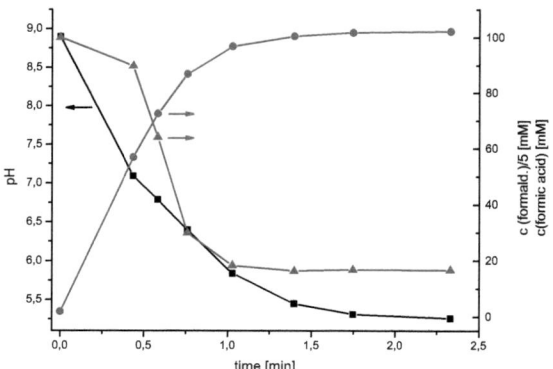

Figure 4.14: Variation of pH (black squares) and calculated amount of formic acid (gray circles) for the reaction of 0.5 M formaldehyde (gray triangles) in 0.1 M K_2HPO_4 at 200 °C.

tration can be calculated with the law of mass action, taking into account the self disproportionation of water and the equilibria of the different phosphate species and formic acid (with the dissociation constants $K_w = 10^{-14}$, $pK_1 = 2.15$, $pK_2 = 7.20$ and $pK_3 = 12.35$ for phosphoric acid [114] and $pK = 3.76$ for formic acid [115]). The calculated concentration is also plotted. Large amounts of acid are created. The maximum concentration of formic acid reaches 100 mM, which is one fifth of the formaldehyde initially present. The Cannizzaro reaction thus accounts for 40% of the formaldehyde consumption. This loss is much higher than in the conventional formose reaction at moderate temperatures. Due to the low salt concentration, the buffer capacity is quickly exhausted. When the formation of carbohydrates finally takes place, the solution is neutral or even slightly acidic.

4.3.5 High Buffer Concentration

The change in basicity might indeed stop the formose reaction. To conduct the reaction under stable pH conditions, the salt concentration was raised to 0.5 M. A sodium carbonate buffer, consisting of equal proportions Na_2CO_3 and $NaHCO_3$, was employed. With these modified conditions the pH remained in the alkaline range

4.3 Hydrothermal Formose Reaction

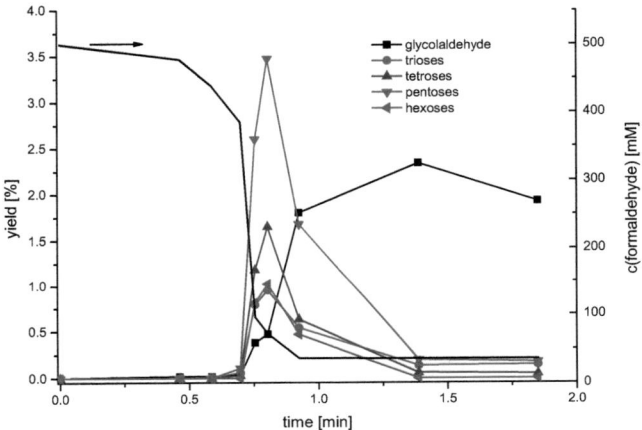

Figure 4.15: Kinetics of formaldehyde consumption and product formation of a 0.5 M formaldehyde solution in 0.5 M carbonate buffer at 150 °C.

even for prolonged reaction times. Performing the reaction at 200 °C resulted in a dark brown solution already after less than half a minute. Contrary to the reactions with low buffer capacity, the solution at high pH did not become turbid, even for prolonged reaction times. It appears that no carbonization occurs in alkaline media.

Due to the fast reaction, the temperature was lowered to 150 °C. Here, kinetics is comparable to the system with 0.1 M K_2HPO_4 at 200 °C. However, the induction period and the subsequent formation of carbohydrates are clearly more distinct (see figure 4.15). The formation of carbohydrates occurs quite abruptly. It takes only 6 s from the beginning of their formation to maximum yield. Again, the shorter carbohydrates account for the major part, though the relative fraction of hexoses is more pronounced than at higher temperatures (compare with figure 4.12). The product distribution is slightly altered. In the strong alkaline media, pentoses are formed preferentially. When the buffer concentration was low and the final pH slightly acidic, pentoses, tetroses and trioses were formed in equal amounts. Besides the very fast formation of carbohydrates, their decomposition is highly accelerated in alkaline media. This is probably the reason why the overall yield was not improved by a higher buffer capacity. Under those conditions it reaches only 7.6%. It turns out that the lower buffer capacity might even have been advantageous. At the end of the reaction the pH turned neutral, and sugars are more stable under these conditions than at high pH.

4.3.6 Carbohydrate Selectivity

It is an interesting question whether the nature of the added salt or catalyst had some effect on the product selectivity. Table 4.2 shows the maximum yield with different salts. Salts are ordered according to increasing basicity. The total yield nicely indicates that the starting pH and not the ion catalysis is the decisive factor. When working under hydrothermal conditions the yield is lower compared to moderate temperatures, but less harsh conditions, e.g. a significantly lower alkaline pH, are necessary to induce the formose reaction. Among the hydrothermal reactions, using a simple 0.1 M carbonate buffer resulted in the highest relative yield. Concentration and type of buffer are optimal, as the solution is initially alkaline, allowing the formose reaction to proceed. When formaldehyde was consumed and the maximum of carbohydrate yield reached, the solution became neutral. Here, sugars exhibit the highest stability.

Taking a look at the various reactions performed at 200 °C no difference in selectivity towards a special sugar became visible. The relative amounts of carbohydrates with different carbon number are roughly identical. As discussed earlier, the most striking difference is the virtually complete absence of hexoses at 200 °C and the preferential formation of shorter carbohydrates.

Two different alditols, erythritol and threitol, originate from the post reaction reduction of the tetroses. Both are formed in approximately equal amounts at 200 °C as well as at 60 °C. This is different regarding pentoses. Here, three sugar alcohols are identified. Under the assumption that all possible stereoisomers of pentoses are synthesized with the same probability, one should obtain 50% arabinitol and 25% of ribitol and xylitol each. Indeed, sugars yielding arabinitol accounted for roughly half of the (linear) pentoses. However, we found a higher concentration of ribitol compared to xylitol under hydrothermal conditions. This selectivity is reversed at 60 °C.

4.3 Hydrothermal Formose Reaction

Table 4.2: Maximum yield of different carbohydrates at hydrothermal and moderate temperatures of a 0.5 M formaldehyde solution with different salts; concentration of salt was 0.1 M except for $Ca(OH)_2$ with 0.05 M.

	yield at maximum [%]				
	200 °C, 100 bar				60 °C
	$Ca(OAc)_2$	$NaHCO_3$	K_2HPO_4	$NaHCO_3$/ Na_2CO_3	$Ca(OH)_2$
glycolaldehyde	0.95	0.80	1.41	1.20	0.55
trioses	0.56	1.41	2.40	2.87	1.73
tetroses	0.87	1.67	2.99	3.19	3.15
as erythritol	0.51	0.82	1.54	1.53	1.54
as threitol	0.37	0.85	1.45	1.66	1.62
pentoses	0.90	1.46	2.48	2.94	9.17
as ribitol	0.25	0.41	0.73	0.82	1.65
as arabinitol	0.49	0.69	1.22	1.38	4.27
as xylitol	0.16	0.35	0.52	0.74	3.25
hexoses	0.16	0.16	0.30	0.29	11.79
total	**3.44**	**5.50**	**9.58**	**10.48**	**26.50**

We can speculate that this is an effect of different reactivities of the pentoses. Figure 4.16 shows their fraction against reaction time for the hydrothermal reaction in 0.1 M K_2HPO_4 and the reaction at 60 °C in 0.05 M $Ca(OH)_2$. In the early stages ribitol dominates over xylitol in both cases. Under $Ca(OH)_2$ catalysis this is reversed very rapidly. Under hydrothermal conditions however, this only occurs for very prolonged reaction times, when the overall yield of sugars is already very low. It was observed that in a neutral solution at 100 °C, ribose decomposes faster than xylose, correlating with the percentage of free aldehyde in solution [101]. Although ribose seems to be formed preferentially, its reactivity is higher. At 60 °C a possible sink for ribose is the continued reaction to hexoses, which is negligible under hydrothermal conditions. Therefore, the fraction of ribitol does not decrease until the maximum yield of carbohydrates has been reached.

Chapter 4 Prebiotic Carbohydrate Synthesis

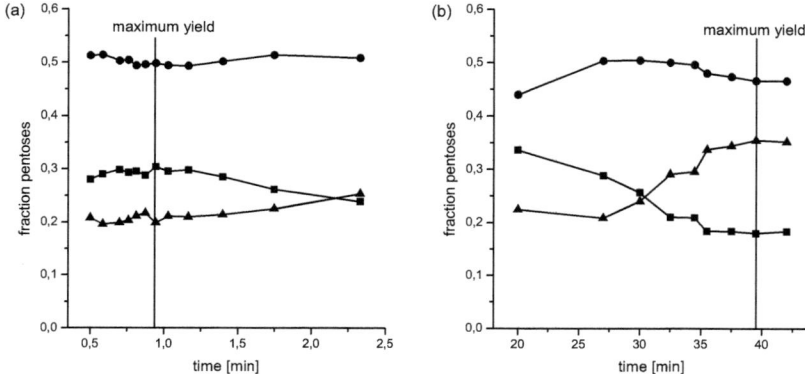

Figure 4.16: Fraction of pentoses showing arabinitol (circles), ribitol (squares) and xylitol (triangles) for a) the hydrothermal reaction in 0.1 M K_2HPO_4 and b) the reaction at 60 °C in 0.05 M $Ca(OH)_2$.

4.3.7 Effect of Temperature

The effect of temperature was investigated in the range of 125 °C to 200 °C in 0.1 M K_2HPO_4 (figure 4.17). Here, 1 mol% glycolaldehyde was added to reduce the reaction time, as it is difficult to handle extended reaction times with the employed flow setup. In particular we checked whether more hexoses would be synthesized at lower temperature. Surprisingly this was not the case, at least not in significant terms and in the analyzed temperature range. Chromatograms at lower temperatures revealed that less side products are formed. For instance, the detectable amount of 2-deoxyribitol decreased. However, total yields were not significantly affected.

4.3.8 Sugar Stabilization

The characteristics of the formose reaction, induced by addition of K_2HPO_4 to the formaldehyde solution, are equally valid for other salts. Only the overall yield is influenced by the nature of the additive, but not the relative product distribution. However, when the additive contains reactive sites, other reaction pathways are possible, which could push the formose reaction towards a certain product. Since the fast

4.3 Hydrothermal Formose Reaction

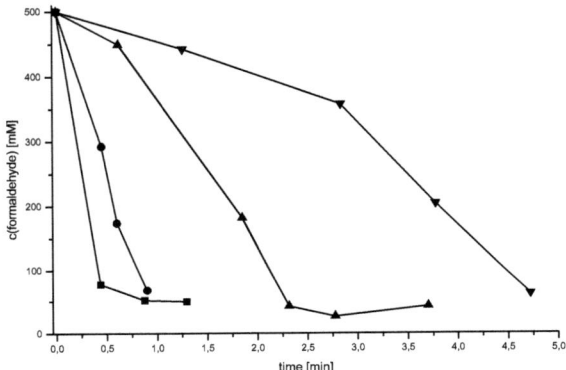

Figure 4.17: Consumption of formaldehyde in 0.1 M K_2HPO_4 with 1 mol% GA at 200 °C (squares), 175 °C (circles), 150 °C (upward triangles) and 125 °C (downward triangles).

decomposition is the main cause for the low yield, the reaction can only be improved by either stabilizing or trapping the products.

Borate minerals were shown to stabilize ribose and other carbohydrates by complexation [116]. When the reaction was performed in 0.125 M $Na[B(OH)_4]$ or in a 0.125 M borate buffer prepared with borax, a fast conversion of formaldehyde was observed (figure 4.18), but NMR spectra revealed that only the Cannizzaro reaction took place. Consumption of formaldehyde was faster in solutions exhibiting higher pH. Even when 1 mol% glycolaldehyde was added as initiator, only marginal amounts of carbohydrates were formed which furthermore decomposed quite rapidly. Also increasing borate concentration to 0.5 M and accordingly lowering temperature to 130 °C did not result in significant carbohydrate formation. Obviously borate is an efficient catalyst of formaldehyde disproportionation under hydrothermal conditions and is unable to stabilize carbohydrates at elevated temperatures.

Another possible means of stabilization, trapping of products with amines, was investigated. Due to its prebiotic relevance, the reaction was performed in the presence of 0.1 M adenine (its solubilization was achieved by heating to 100 °C). No additional salts were added. Formaldehyde consumption as well as a colour change were observed. However, no free carbohydrates could be detected and NMR spectra

Chapter 4 Prebiotic Carbohydrate Synthesis

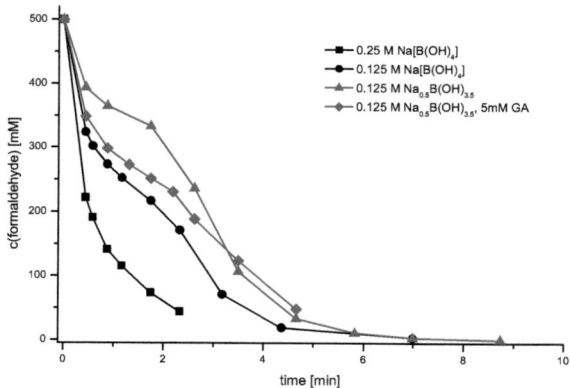

Figure 4.18: Conversion of 0.5 M formaldehyde at 200 °C in presence of borate (black curves) and borate buffer (gray curves).

did not show any carbohydrate signals in the region of 3.5–4.5 ppm (figure 4.19). Clearly, the presence of amines blocks the formose reaction and directs it towards other pathways.

When zinc prolate was used as additive in the formose reaction, a very rapid consumption of formaldehyde took place (see figure 4.6, page 56). This salt was shown to be an efficient catalyst for aqueous aldol reactions [109]. At the high temperatures used in our experiments, however, conversion levels of proline were similar to those of formaldehyde, indicating an incorporation of the amino acid into the products. Again, no carbohydrates could be detected.

In a last experiment adenine was not reacted solely with formaldehyde, but in the presence of a 0.5 M carbonate buffer. At high pH adenine is present as base, which might increase reactivity. The same settings as in the experiment without adenine (150 °C, 100 bar) were chosen. Kinetics and colour change were similar in both cases (see figure 4.15, page 67). Likewise, the concentration of free carbohydrates was comparable. Interestingly, a white precipitate formed slowly after reaction. After cooling in an ice bath, the solid was filtered and an NMR spectra recorded in DMSO (figure 4.20). It shows peaks around 8 ppm, which are typical for the adenine scaffold. However, at least 4 peaks are obtained in this region, instead of the expected 2. The

4.3 Hydrothermal Formose Reaction

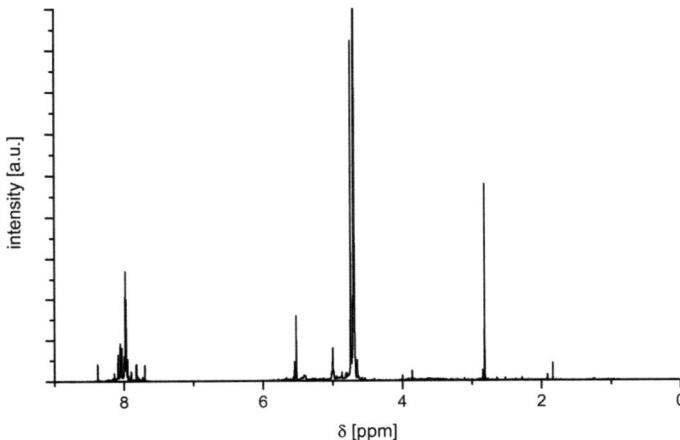

Figure 4.19: ^{1}H-NMR of the reaction of 0.5 M formaldehyde with 0.1 M adenine at 200 °C for 5.2 min lacks of peaks corresponding to carbohydrates; recorded in D_2O.

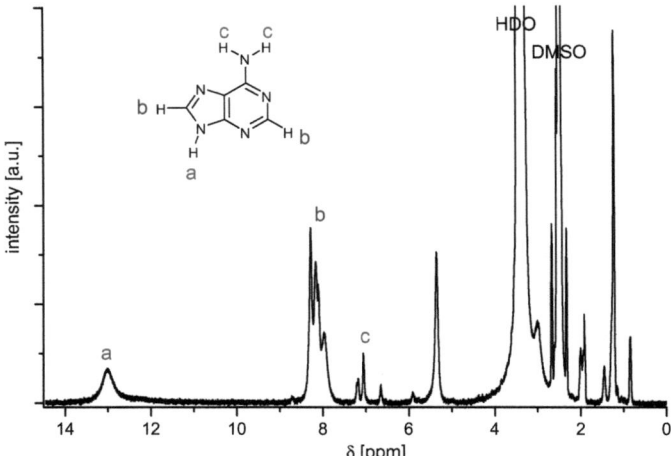

Figure 4.20: ^{1}H-NMR of the precipitate from the formose reaction in the presence of adenine, reaction time 1.2 min; recorded in DMSO.

73

Chapter 4 Prebiotic Carbohydrate Synthesis

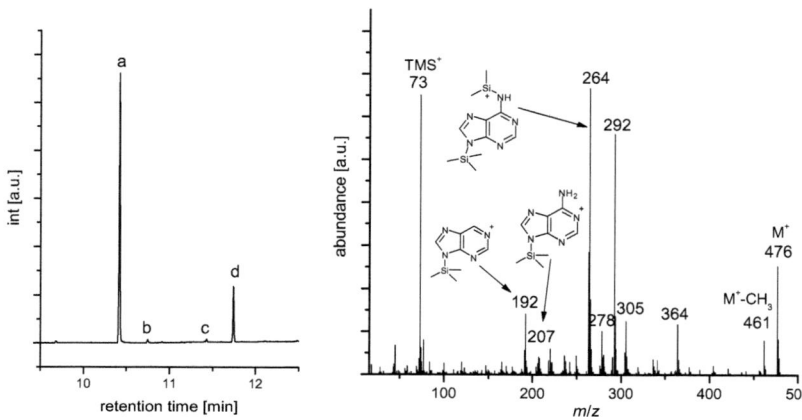

Figure 4.21: GC of the silylated precipitate.

Figure 4.22: MS of the unknown compound d in the precipitate.

intensity of signals from protons of the amino groups is less than expected, which might be due to an exchange with residual water.

Additional signals appear, notably one around 5.5 ppm. It could originate from a double bond or a hemiaminal, obtained after the addition of formaldehyde [117]. Other signals between 1–2 ppm are also present, indicating the presence of alkane protons. No carbohydrates are present, based on the absence of signals between 3.5–4.5 ppm.

Since these findings are difficult to explain solely with NMR measurements, the white precipitate was silylated with BSTFA and a GC recorded (figure 4.21). It shows two main peaks (a, d) and two smaller ones (b, c). Compound a was identified as residual adenine, which was silylated twice. The molecular ion was clearly visible in its mass spectrum. Compound b is also adenine, now bearing three TMS-groups. Here, each acidic proton was replaced by the silyl moiety.

Compounds c and d also exhibit a mass shift of 72 for the largest fragment, which is the shift expected for the incorporation of one TMS-group. The mass spectrum of d is shown in figure 4.22. The signal at $m/z = 476$ is most probably the molecular ion M^+. A loss of 15 is common among trimethylsilylated compounds, resulting

Table 4.3: Comparison of formose reaction characteristics under hydrothermal and moderate temperatures

high temperature and pressure	moderate temperature
catalytically active ions have minor effects	catalytically active ions necessary
initiator improves yield	initiator does not influence yield
induction period even without initiator quite short	initiator drastically reduces induction period
nature of initiator does not influence outcome of reaction	nature of initiator does not influence outcome of reaction
shorter carbohydrates are formed, rarely hexoses	main products are hexoses and pentoses
selectivity pentoses: arabinitol > ribitol > xylitol	selectivity pentoses: arabinitol > xylitol > ribitol
lower yield	higher yield

from the elimination of a methyl group. Interestingly, the spectrum for m/z-values up to 278 is very similar to that of *di*-TMS-adenine. We can therefore conclude that the adenine scaffold is still retained in the unknown compound. However, the identification of the additional moiety still requires further investigation.

The exact number of silyl groups could for instance be determined using a different silylation agent, e.g. incorporating a dimethyl *tert*-butyl silyl moiety instead of the TMS one. The use of ^{13}C-formaldehyde would provide the number of formaldehyde molecules that react with adenine. Furthermore, performing the reaction in D_2O would probably lead to the exchange of protons with deuterium. In another experiment with labeled glycine it will be shown that even non-acidic protons are rapidly exchanged at high temperatures (see section 6.2.1, page 94). The structure of the unknown compound has to be elucidated in further studies. The formose reaction is thus, despite its antiquity, highly interesting and surprising results can be expected in the future.

4.4 Summary

The feasibility of the formose reaction under hydrothermal conditions was analyzed. In contrast to the counterpart at ambient temperature, less demanding conditions are required. Catalytically active species are not essential and only have minor effects regarding the yield. A slightly alkaline solution is sufficient to induce the reaction. In fact, depending on buffer capacity, the pH can even drop to slightly acidic at the end of the reaction. A high pH throughout the whole reaction is in fact disadvantageous, as product decomposition is favoured above a certain pH.

No selectivity towards a particular product could be detected. However, considerable differences in the outcome of the reaction with respect to moderate temperature were found. Hexoses, being the main product at low temperature, are only formed in negligible amounts. Regarding the distribution of pentoses, a reversed selectivity towards carbohydrates yielding ribitol and xylitol was found. Besides the true condensation products of formaldehyde, also deoxysugars were detected. Hydrothermal conditions seem to facilitate alternative reaction pathways [118]. The low yield caused by rapid decomposition is a key issue. Attempts to stabilize or trap the products under hydrothermal conditions have been unsuccessful so far. These findings do not provide convincing evidence that the formose reaction played a role in the prebiotic formation of carbohydrates. Still, they nicely demonstrate the extent hydrothermal reactions can differ from moderate temperature chemistry. The aforementioned characteristics of the formose reaction under hydrothermal and moderate temperatures are summarized in table 4.3.

Chapter 5

Transfer Hydrogenation of Levulinic Acid

5.1 Background

Being the main component of biomass, carbohydrates are the most promising starting compound for replacing fossil resources and moving towards a sustainable chemistry. Apart from gasification, which yields small molecules, it is energetically advantageous not to destroy the complete molecule, but to break down the carbohydrates step by step. This can be achieved by current biotechnology processes (see section 2.1.3 on page 13). Ethanol, lactic acid, succinic acid or citric acid can be obtained by fermentation [11]. A very promising chemical pathway is the dehydration of carbohydrates to hydroxymethylfurfural. This can be obtained from fructose by simple heating in an acidic aqueous environment with continuous extraction of the product [119]. Under non-aqueous conditions in ionic liquids and with metal ion catalysts like $CrCl_2$, this molecule can also be synthesized from glucose with fairly high yield [120]. In water, especially under acidic conditions, hydroxymethylfurfural rehydrates and reacts further to levulinic and formic acid (figure 5.1) [121].

Levulinic acid itself is an interesting building block and can be converted to a variety of useful compounds [122]. With respect to green chemistry, γ-valerolactone is potentially of highest significance. One remarkable characteristic is its applicability as sustainable liquid [123]. It is nontoxic with an LD_{50} significantly higher than ethanol, has favourable solvent characteristics and exhibits both a low melting

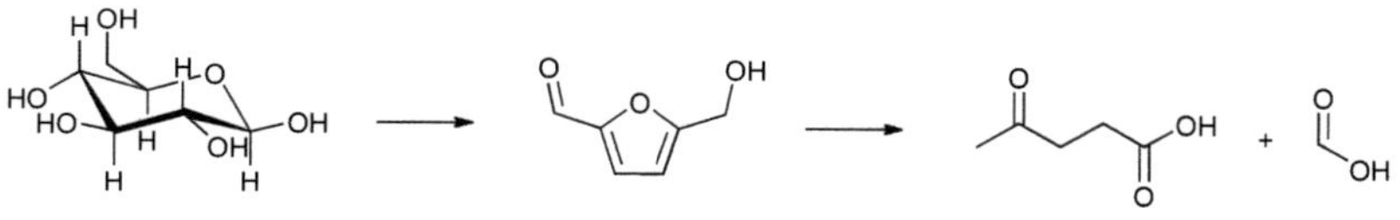

Chapter 5 Transfer Hydrogenation of Levulinic Acid

Figure 5.1: Decomposition of glucose to levulinic and formic acid with intermediate formation of hydroxymethylfurfural.

point as well as a high boiling point and very low vapour pressure, even at elevated temperatures. It is also biodegradable and thus can be considered a green solvent. In addition to replacing toxic organic solvents, γ-valerolactone is useful as an energy storage molecule and can be directly employed as an additive for liquid transportation fuels. Also, alkanes can be obtained by further hydrogenation[124].

γ-Valerolactone can be synthesized by hydrogenation and subsequent cyclisation of levulinic acid. Both heterogeneous [125–128] and homogeneous catalysts [124, 128, 129] have been evaluated so far. However, both systems employ noble and heavy metals, in particular palladium and ruthenium, which are neither sustainable nor low-priced. Since an equimolar amount of formic acid is generated during production of levulinic acid from carbohydrates, it is convenient to use formic acid as reducing agent in a transfer hydrogenation (figure 5.2). In this way, the side product of a first reaction is used as a hydrogen source for the second, and a high atom and energy efficiency is achieved.

To make this process even more environmentally friendly, the heavy metal catalysts have to be replaced. The reduction of aldehydes with NaCOOH has been shown to work without catalyst, though very high temperatures were necessary [130]. This result was the motivation for studying the transfer hydrogenation of the even more unreactive ketone using formic acid under green conditions in water [131].

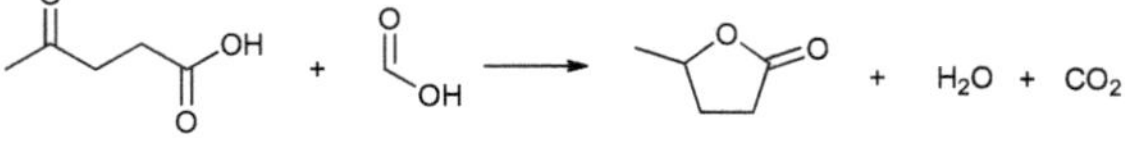

Figure 5.2: Transfer hydrogenation of levulinic acid with formic acid as hydrogen donor.

5.2 Results

5.2.1 Salt Effects

Since levulinic acid and formic acid are produced from carbohydrates in aqueous phase, the transfer hydrogenation was performed in diluted solution. This is practical as separation of the two acids from water would be highly energy consuming. Keeping this in mind, 0.1 M levulinic acid was used with a slight excess of formic acid (0.15 M). To preliminary test, whether synthesis of γ-valerolactone under hydrothermal conditions is feasible or not, the mixture of levulinic and formic acid was simply heated in an autoclave at 220 °C for 12 h, resulting in a yield of only 1.0% (analyzed by HPLC). By addition of Pd on activated alumina (12 mol% Pd with respect to levulinic acid), the yield increased to 29.0%. However, this was accompanied by a significant occurrence of further reduction towards pentanoic acid, as proven by GC-MS after silylation of the sample. Furthermore, Pd not only catalyzed the desired hydrogen transfer, but also the decomposition of formic acid to H_2 and CO_2. This decomposition, although at much lower rate, also occurs in high-temperature water without any additives [51]. This side reaction, together with the demand for a cheap, sustainable catalytic scheme, makes the use of Pd@Al_2O_3 obsolete. Still, it was useful in this system to obtain a benchmark value for the possible conversion range. By using catalysts based on ruthenium, much higher yields have been reported [128, 129], but Ru has the same disadvantages as Pd concerning sustainability.

In hydrothermal reactions, even simple salts may activate the water and can therefore act as catalysts [132]. As cations might activate the carboxy group by coordination, a 0.5 M salt solution was used as solvent instead of pure water under the same reaction conditions. Different chlorides (KCl, $CeCl_3$, $CoCl_2$, $SrCl_2$, $CuCl_2$ and $ZnCl_2$) were tested, but none showed a significant activity for this reaction, not even the employed transition metals (table 5.1).

Taking a closer look at the different halides, one can notice some effects following the Hofmeister series (see table 2.1 on page 23). Potassium fluoride increases the yield compared to the pure aqueous solution, whereas the other halides do not show any improvement, while a slight decrease from chloride to the more chaotropic iodide is observed. A remarkable acceleration is also found when Na_2SO_4 is added. Fur-

Chapter 5 Transfer Hydrogenation of Levulinic Acid

Table 5.1: Effect of salt addition (0.5 M) on the yield of γ-valerolactone

additive	yield [%]	additive	yield [%]
no salt	1.0	KF	11.3
Pd@Al$_2$O$_3$	29.0	KCl	1.4
KCl	1.4	KBr	1.2
CeCl$_3$	1.4	KI	0.7
CoCl$_2$	1.0	Na$_2$SO$_4$	11.0
SrCl$_2$	1.1	Na$_2$SO$_3$	0.0
CuCl$_2$	0.0	K$_2$CO$_3$	0.0
ZnCl$_2$	0.1	KH$_2$PO$_4$	3.4
		K$_2$HPO$_4$	0.7
		K$_3$PO$_4$	0.0

thermore, KH$_2$PO$_4$ increases the yield, whereas the more basic phosphates do not show any catalytic activity, even inhibiting the reaction. Finally, other basic salts like Na$_2$SO$_3$ and K$_2$CO$_3$ prevent any reaction. This makes pH a potential key factor in this reaction.

Therefore, we had to investigate the pH dependence of the transfer hydrogenation without additional salt to distinguish salt and pH effects. As both educts are acids, a variety of reaction pathways can be considered. Whereas formic acid releases CO$_2$ and H$_2$ upon decomposition, its anion formally transfers H$^-$. Thus the reducing character of formic acid is significantly different from that of formates. To perform the reaction at different pH, variable amounts of either hydrochloric acid or potassium hydroxide were added to the mixture of levulinic and formic acid. The concentration of HCl was 0.01 M or 0.1 M in the final solution, whereas 0.075 M, 0.15 M, 0.2 M or 0.3 M of KOH was added to shift the pH to higher values. In the case of base addition, pH is buffered by formic acid ($pK_a = 3.74$) and levulinic acid ($pK_a = 4.59$). Since during the reaction acid is consumed or may decompose, the pH should increase throughout reaction. Indeed, this is validated by the experiment (figure 5.3). The pH changes are largest in the alkaline region, presumably due to the missing buffer capacity.

Furthermore, the yield strongly depends on the pH. At very acidic pH, where both reactants are protonated and uncharged, the yield is very low. With increasing pH,

5.2 Results

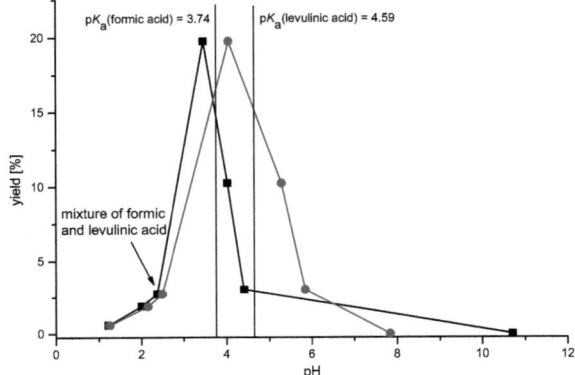

Figure 5.3: The dependence of γ-valerolactone yield upon pH, measured before (squares) and after (circles) the experiment.

the yield drastically increases, reaching a maximum at a pH around the pK_a of formic acid. Further increasing the pH leads again to a decreasing yield. Levulinic acid now becomes deprotonated, which seems to inhibit the reaction. Obviously, the reaction is fastest for levulinic acid in its neutral and formic acid in its anionic form. This leads to a rather sharp maximum of optimum pH for product generation in water with a formal transfer of H⁻ from formate. So by simple addition of some base to the mixture of levulinic and formic acid, the yield can be drastically optimized. In media that are too basic, lactonization is not expected to occur. However, GC-MS did not reveal any other low molecular weight compounds besides levulinic acid (especially not γ-hydroxyvaleric acid), indicating that no reaction at all seems to occur at high pH. Another potential reason for the low yield under acidic conditions is the fact that addition of HCl promotes the autodecomposition of formic acid at high temperatures (figure 5.4), which is — in addition to the reaction with levulinic acid — a second reaction channel.

Knowing the pH sensitivity of the transfer hydrogenation of levulinic acid with formic acid, the initial experiments on the influence of pure salts on the reaction can be reanalyzed. The pH probably accounts for the acceleration pattern of the different potassium phosphates. It is now also evident why very basic salts like Na_2SO_3 or

Chapter 5 Transfer Hydrogenation of Levulinic Acid

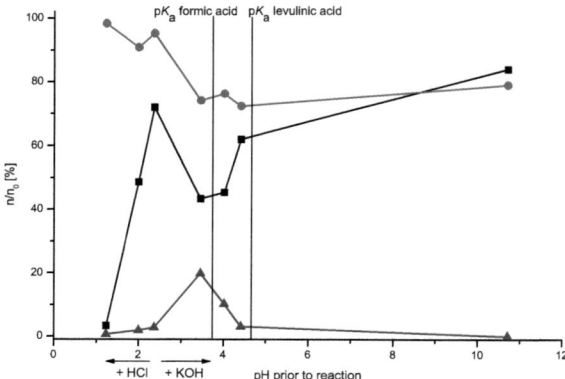

Figure 5.4: Amount of levulinic acid (circles), formic acid (squares) and the product γ-valerolactone (triangles) versus pH prior to reaction.

Na_2CO_3 inhibit any reaction. In contrast, slightly basic salts could shift the pH in the narrow optimum range. This might be the case for KF and Na_2SO_4.

5.2.2 Dissociation at High Temperature

The catalytic activity of sulfate, which is a quite neutral salt ($pK_a = 1.99$ for the deprotonation of HSO_4^- under ambient conditions) is still surprising at first sight, whereas fluoride is more basic ($pK_a = 3.17$ under ambient conditions). Despite the difference in K_a of one order of magnitude, the two different salts have a similar effect on the reaction. When determining the basicity of the tested salts, one should bear in mind that the pK_a-values at elevated temperatures can be quite different from those under ambient conditions. So to really relate the effect to pH, we have to compare the dissociation constants at the reaction temperature of 220 °C. The temperature dependence of the pK_a-values can be expressed by the van't Hoff equation

$$\frac{d \ln K_a}{dT} = \frac{\Delta H^0}{RT^2} \tag{5.1}$$

5.2 Results

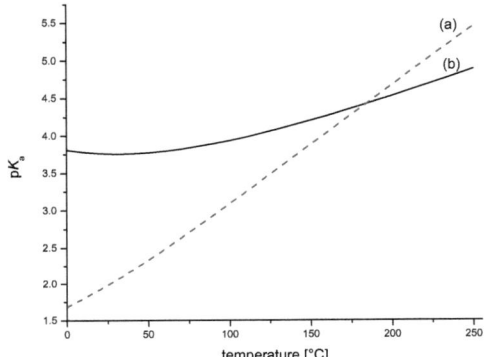

Figure 5.5: Variation of pK_a of HSO_4^- (a) and formic acid (b) with increasing temperature, calculated from thermodynamic data.

where ΔH^0 is the standard molar enthalpy change for the dissociation of the acid. Integration, neglecting the increased pressure at high temperatures and the temperature dependence of the heat capacity change Δc_p^0, leads to

$$K_a(T) = -\frac{\Delta G^0}{T^0} + \Delta H^0 \cdot \left(\frac{1}{T^0} - \frac{1}{T}\right) + \Delta c_p^0 \left(\frac{T^0}{T} - 1 + \ln\left(\frac{T}{T^0}\right)\right) \quad (5.2)$$

with T^0 being the reference temperature ($T^0 = 298.15$ K). With data from literature for formic acid ($\Delta G^0 = 21.45$ kJ · mol^{-1}, $\Delta H^0 = 1.03$ kJ · mol^{-1} and $\Delta c_p^0 = -175$ J · K^{-1} · mol^{-1} [133]) and HSO_4^- ($\Delta G^0 = 11.342$ kJ · mol^{-1}, $\Delta H^0 = -22.4$ kJ · mol^{-1} and $\Delta c_p^0 = -258$ J · K^{-1} · mol^{-1} [114]) the temperature dependence of the pK_a of formic acid and HSO_4^- can be calculated (figure 5.5).

Like for most simple acids, the dissociation constant of formic acid decreases only slightly with increasing temperature. Even though the pressure dependence was not taken into account, the calculated pK_a variation is in good agreement with experimental values for both formic acid [115, 133] and hydrogensulfate [134]. On the other hand, the dissociation constant of HSO_4^- decreases very strongly and even crosses that of formic acid at elevated temperatures. The reason is the very high enthalpy change for the dissociations of HSO_4^-, causing K_a to be highly affected by

Chapter 5 Transfer Hydrogenation of Levulinic Acid

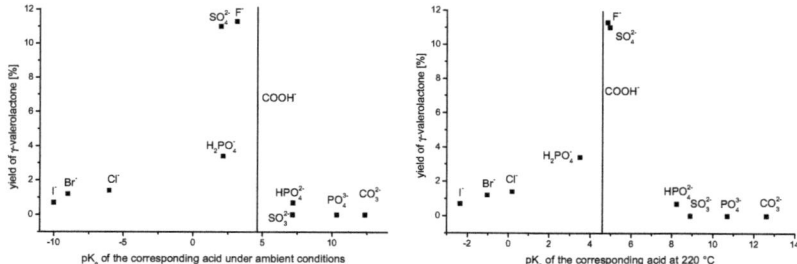

Figure 5.6: Yield versus the pK_a of the corresponding acids of the tested salts under ambient conditions and at 220 °C or 200 °C for the halides respectively.

temperature. At the high reaction temperatures in the experiments sulfate can in fact act like a weak base with a strength comparable to carboxylic acid salts and becomes protonated to a significant extent while deprotonating formic acid. We speculate that this also causes the high yield of γ-valerolactone in the presence of this salt.

For the other tested salts the pK_a values were also calculated for a temperature of 220 °C [114]. In the case of halides, data for 200 °C were used [135]. One can see clear differences compared to the values at ambient conditions, which cannot explain the action of the different salts in contrast to the values at high temperatures (figure 5.6). Here the pK_a of hydrogen fluoride is quite near to the value of HSO_4^-, which results in a similar yield of product by adding KF. The difference in dissociation constant is, however, less pronounced than for HSO_4^-. In a green, sustainable approach, KF cannot be recommended due to its toxicity. On the other hand there is no concern in using simple sulfates. In any case, the KF example allows the conclusion that the observed effects of salt addition are not attributed to Hofmeister effects, but rather to their influence on pH at elevated temperatures.

5.2.3 Optimization of Reaction Conditions

To gain further insight into the effect of added Na_2SO_4, standard mixtures of levulinic and formic acid were reacted in the presence of different concentrations of Na_2SO_4. Increasing the salt concentration increases the yield, until a plateau is reached around 0.1 M of salt (figure 5.7). Interestingly, the yield increases roughly with the logarithm

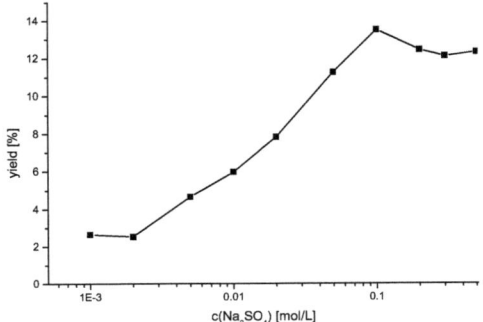

Figure 5.7: Catalytic effect of Na_2SO_4 on the transfer hydrogenation.

of added salt concentration.

For further optimization of the reaction conditions the temperature dependence of the reaction was screened within an 8 mL tubular flow reactor for a fixed flow of 0.4 mL min^{-1} and a pressure of 200 bar, corresponding to a reaction time between 18.1 min (at 175 °C) and 14.7 min (at 300 °C) (figure 5.8). With its continuous product stream, such a setup is certainly more adapted to industrial processes. In these experiments 0.125 M Na_2SO_4 was used, as the reaction performed in autoclaves

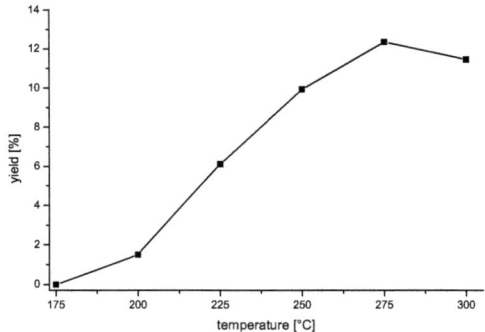

Figure 5.8: Variation of yield versus reaction temperature for a residence time in the range of 18.1 min (at 175 °C) and 14.7 min (at 300 °C) with 0.125 M Na_2SO_4 added.

gave satisfactory yields at that concentration. Furthermore, a low concentration ensures complete solubility, which is important in order to not block the reactor. The solubility of Na_2SO_4 strongly decreases around 300 °C (see figure 2.10 on page 27). This experiment shows that a sufficiently high temperature to activate the formic acid is necessary to start the reaction. Product yield under those conditions passed through a maximum, as a temperature above 275 °C again lowers the yield. It is assumed that the spontaneous decomposition of formic acid is too fast in this range. Hydrogen, formed by this process, cannot reduce levulinic acid. It is the direct hydrogen transfer which creates the γ-valerolactone. This was validated by another experiment, where 0.1 M levulinic acid in 0.5 M Na_2SO_4 was reacted with hydrogen (5 bar, 10 eq.) at 220 °C for 16 h. Here, no γ-valerolactone was formed.

Comparing the overall yield from this experiment with the reaction performed in autoclaves (see table 5.1 on page 80), one notices the decreased yield at the reaction temperature of 220 °C. This is due to the fact that in the tubular flow reactor the residence time is less than 20 min, as compared to the batch mode, where it is 12 h. The short reaction time therefore results in incomplete conversion. Higher yields are expected when pumping the reaction mixture into the reactor for a second time. The highest potential for an increased yield is given when the γ-valerolactone is extracted prior to re-reacting the unconverted starting educts, but such optimization by chemical engineering is out of the scope of the presented work. Selective extraction can be carried out for example with supercritical CO_2 [136].

5.3 Summary

The synthesis of γ-valerolactone can be performed under hydrothermal conditions, although improvement of yield is necessary if this process is to be adapted for large scale throughputs. Nevertheless, the reaction is highly atom and energy efficient. In regard to the whole process starting from carbohydrates, no other reactants, such as an external hydrogen supply, are required, and even side products are converted. γ-Valerolactone has a very high potential as solvent and fuel molecule that is compatible with the requirements of green chemistry. The presented synthesis does not depend on conventional heavy metal catalysts, which is another sustainable feature.

Optimization of pH allows an increase in the overall yield, but the reaction is also highly affected by the presence of salts. Even very simple salts, like Na_2SO_4, can act as catalysts and drastically improve the yield. This "salt catalysis" cannot be attributed to Hofmeister effects but is probably purely related to their influence on pH. A special property of Na_2SO_4 was used to optimize the reaction. Sulfates in high temperature water are "switchable" bases and become more and more basic with increasing temperature. This temperature dependence is more pronounced for sulfate than for other anions. It can be exploited for easy preparation of reaction solutions that are neutral and thus easy to handle under ambient conditions, but drastically change their properties at high temperature.

Such a temperature switch may be used to substitute conventional bases that are traditionally employed as catalysts. They require neutralization after reaction and thus create waste salt. Furthermore, a temperature-switchable base will become neutral when cooling to ambient conditions after the reaction and can therefore be reused. Hence, hydrothermal synthesis is a means of preventing waste. This adds another sustainable feature to the high atom economy of the presented synthesis.

Chapter 6

Hydrothermal Biomass Valorization

6.1 Alkaline Digestion

6.1.1 Conversion of Glucose

As previously demonstrated with the example of conversion to levulinic and formic acid and subsequent transfer hydrogenation, the chemistry of carbohydrates is very versatile. Another promising pathway to convert sugars to useful chemicals is their reaction to lactic acid. Especially under alkaline conditions a high yield of lactic acid can be achieved. Additionally, the formation of deoxyhexonic acids and other polyhydroxy acids takes place [137]. The fact that the pH drops when carbohydrates are carbonized was also motivation to study the formation of acids under hydrothermal reaction conditions in more detail.

To prevent carbonization, experiments were performed at 130 °C. A 0.7 M glucose solution was heated for one day with some basic amines (histidine hydrochloride, piperidine and lysine). First, a catalytic effect was tested and a low concentration of 4 w% additive with respect to glucose chosen. In all experiments a final pH around 4 was reached. To evaluate the amount of acidic groups, the solution was titrated with NaOH. The result is shown in figure 6.1.

Despite the same pH of the hydrothermally treated solutions, the titration curve reveals that when no additive was present only marginal amounts of acids were formed. When additional bases were present, significantly more acid is generated. The amount of acidic groups seems to be proportional to the amount of base added. This proves that the action of base is not catalytic, but it is rather a stoichiometric

Chapter 6 Hydrothermal Biomass Valorization

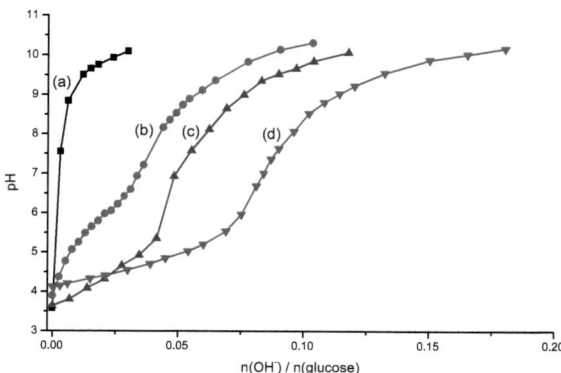

Figure 6.1: Titration of alkaline glucose digestion at 130 °C for 1 d in presence of amines; a) no additive; b) 3.7 mol% histidine hydrochloride; c) 4.9 mol% lysine; d) 11.3 mol% piperidine.

reagent. Glucose decomposes to acids in noteworthy amounts only when the pH is sufficiently high. When it drops below 4, further formation of acids is blocked. Therefore, a stoichiometric concentration of base must be chosen.

To evaluate the maximum amount of base consumed, a 0.5 M glucose solution was reacted with different amounts of NH_3 at 150 °C for 1 d. Here, a higher temperature was chosen compared to the previous experiments in order to accelerate the reaction. It was observed that at sufficiently high base concentration the carbonization reaction was blocked completely, and only a clear, dark brown liquid phase formed. Adding 1.4 eq NH_3 with respect to glucose, a neutral solution was obtained after hydrothermal treatment. Therefore, we can conclude that one mole of glucose can be converted to ≈ 1.4 moles carboxylic acid when ammonia is used.

The dark brown colour of the solution indicates the presence of many decomposition products. To analyze at least the compounds of low molecular weight, gas chromatography was performed after silylation. It showed the formation of many different acids like lactic acid, hydroxyacetic acid, 3-deoxyhexonic acid and many other polyhydroxy acids. Due to the lack of selectivity, no attempts were made to quantify them.

NH_3 not only acts as a base, but can react with sugar and its decomposition products. In fact, we also found alanine and homoserine, products that were absent when other bases were employed instead. Heterocyclic pyrazines were also formed [138], although the yield was low in our experiment (600 mg could be extracted with diethyl ether from 180 g glucose).

Contrary to acidic reaction conditions, levulinic acid is only formed in marginal amounts. The occurence of small organic acids under alkaline conditions shows that glucose is split into smaller fragments. This happens by retro-aldol reactions as a first step [139]. The absence of carbonyl groups and the presence of carboxylic acids in the final products prove that disproportionation by Cannizzaro reactions is occurring. In order to avoid the incorporation of the base into the carbon scaffold, KOH was used for further experiments. Because of the missing selectivity, we also used a less defined feedstock, namely wood from conifers.

6.1.2 Liquefaction of Wood

In section 2.4.1 (page 29) the hydrothermal liquefaction of biomass was illustrated. However, at temperatures around 220 °C it was not possible to liquefy wood completely. 100 g sawdust were reacted in 650 g water with 15–25 g KOH. About 60–50% of the saw dust was converted to soluble species. 15 g KOH was not sufficient, since here the solid residue was highest. 25 g KOH seems to be the optimum amount. After 2 h reaction at 220 °C, the pH was only slightly alkaline (pH = 8.1), indicating a nearly complete consumption of the base.

Wood consists mainly of cellulose, hemicellulose and lignin. We speculate that the carbohydrates can be depolymerized to their monomers, which then degrade to organic acids according to the aforementioned base catalyzed pathways. Lignin, however, is more difficult to decompose. We assume that it mainly stays intact. To check this hypothesis gas chromatography of a silylated sample was performed. The chromatogram is shown in figure 6.2. Important peaks were assigned through a database search of their EI spectra.

The chromatogram clearly indicates the presence of typical carbohydrate decomposition products. Only the major peaks are assigned, showing lactic acid, a deoxyhexonic acid (as its lactone), hydroxyacetic acid, hydroxybutanoic acid and a small peak corresponding to levulinic acid. In addition, many other acids could be de-

Chapter 6 Hydrothermal Biomass Valorization

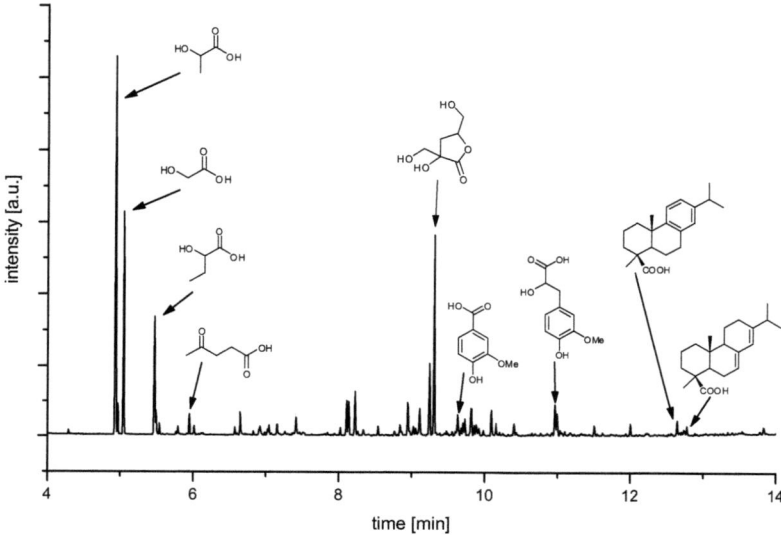

Figure 6.2: Chromatogram of a silylated sample from alkaline wood digestion shows typical decomposition products of carbohydrates and some fragments resulting from lignin, TMS group removed in drawings.

tected. Apart from these compounds lignin related species were also found. They are structurally derived from coniferyl alcohol (compare with figure 2.1, page 14). Also, resin components dehydroabietic acid (at $t_r = 12.6$ min) and abietic acid (at $t_r = 12.9$ min) could be detected. Although the compounds were not quantified, carbohydrate derived acids obviously account for the major part.

Since a broad spectrum of acids is generated with low selectivity, the presented digestion is not useful regarding controlled synthesis. However, the depolymerization and subsequent conversion of carbohydrates into acids are the first two steps in the bacterial production of biogas. Chemically performing these steps could make wood accessible to fermentation and increase methane yield. Therefore, we tested the gas yield of the liquid phase.* Indeed, a fast degradation and high methane yield were

---
*Analysis was carried out by WESSLING Laboratorien GmbH, Oststraße 6, 48341 Altenberge

6.1 Alkaline Digestion

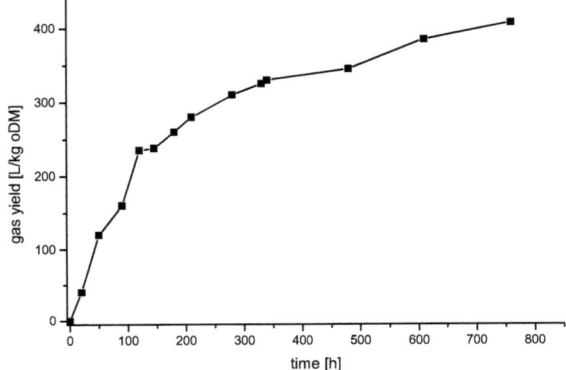

Figure 6.3: Gas yield versus fermentation time for alkaline digested wood.

observed, as shown in figure 6.3. 408 L/kg gas were obtained in 33 days based on the organic dry matter (oDM). It consisted of 65.3 % methane, 25.9 % CO_2 and the rest being mainly N_2. The methane yield was therefore 266.4 L/kg oDM. With carbohydrates as feedstock typically 700–800 L/kg oDM gas with 50% methane are obtained [140], which is slightly higher.

These results have to be regarded under the aspect that wood normally is very hard to digest in a biogas factory. It practically cannot be used as feedstock. Flash pyrolysis was proposed as pretreatment [141]. The pyrolysis oil could also be employed to produce methane. However, it had to be used in conjunction with cellulose to prevent inhibition of the bacteria, and then 371 mL gas were obtained from 1 g pyrolysis oil. In contrast, gas yield of the alkaline treated wood was even slightly higher, and the product was directly suitable for fermentation without added feedstock. Thus the hydrothermal pretreatment is an efficient and energy-saving way to convert wood and probably other low cost plant materials as well into methane.

As a by-product of methanogenesis, CO_2 is formed, which combines with the added base to hydrogen carbonate. For a sustainable process managing the initial base has to be recycled. Substitution by $Ca(OH)_2$ might be a good alternative. Upon heating its carbonate salts loose CO_2 fairly easily, and the initial base could be recycled.

Chapter 6 Hydrothermal Biomass Valorization

6.2 Stability and Decomposition of Glycine

6.2.1 Theory

The formose reaction has demonstrated that carbohydrates can be synthesized under hydrothermal conditions, but also revealed their instability. Amino acids in contrast are more resistant to high temperature. Interestingly, when glycine is heated in a flow reactor, dimers and trimers are formed [142]. Repeated circulation through a hot and cold spot even resulted in the formation of short oligomers with up to six glycine molecules [143]. The main condensation product was, however, not a linear oligomer, but the cyclic dimer 2,5-diketopiperazine.

Kinetics and products of glycine in sub- and supercritical water have recently been studied over a wide range of temperatures [144]. This study clearly demonstrated that the main decomposition pathway of glycine is decarboxylation to methylamine, which is a major observation for most amino acids [145]. In future this might be an alternative way to produce amines from the corresponding amino acids. Less significant in quantitative terms is the dimerization to 2,5-diketopiperazine. In parallel, deamination to glycolic acid also takes place, which can further yield formaldehyde, CO and CO_2. This was the motivation to investigate with sensitive GC measurements, whether different compounds are formed as well, with the aim to elucidate possible reaction networks under hydrothermal conditions.

6.2.2 Results

A 1 w% solution of glycine in pure water was reacted at 180 °C and 250 °C under 100 bar. The conversion of glycine is shown in figure 6.4. This amino acid is quite stable, so that at 180 °C it reaches only 25 % for 7 min reaction time. At 250 °C the kinetics is however quite rapid, and 70 % conversion were obtained after 3 min.

This sample with highest conversion was freeze dried and an NMR spectra recorded (figure 6.5). The three main peaks can be attributed to glycine (a), 2,5-diketopiperazine (b) and methylamine (c). Chemical shifts compare well with literature [144]. Methylamine can be found in the sample despite its volatility, as it is probably fixed as carbonate salt during freeze drying. Furthermore, several smaller peaks appear between 1.5–4 ppm. At low field no peaks can be detected, except for a very tiny

6.2 Stability and Decomposition of Glycine

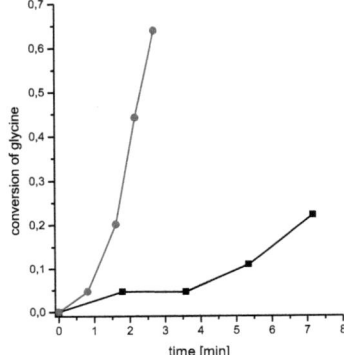
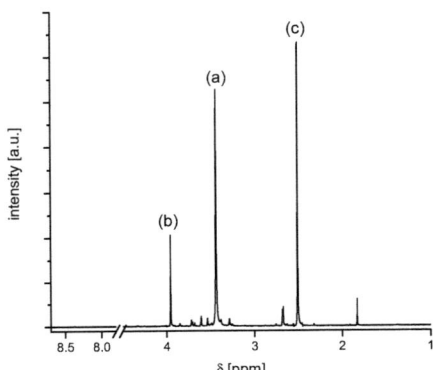

Figure 6.4: Conversion of 1 w% glycin at 180 °C (gray circles) and 250 °C (black squares).

Figure 6.5: NMR of the glycine decomposition experiment at 250 °C shows residual glycine (a), cyclic dimer (b) and methylamine (c).

one in the aromatic region at 8.4 ppm.

As the NMR spectrum indicates, other compounds besides the well described ones can be expected, although only present in small amounts. To identify them, samples from the experiments at 250 °C were silylated after freeze drying and a GC was recorded. Chromatograms are shown in figure 6.6 and clearly reveal that several compounds originate from the decomposition of glycine. These chromatograms display only peaks with a retention time longer than that of the silylation agent BSTFA. Mass scans at earlier times show the presence of methylamine and small amounts of formic and acetic acid. The two dominating signals are residual glycine, either as *di*-TMS (peak *d*) or *tri*-TMS (peak *e*) derivative. In principle only the *tri*-TMS-glycine could be obtained using higher derivatization temperatures. However, we observed that then other peaks started to deteriorate and further signals appeared. To avoid such artifacts a relatively mild temperature (70 °C) was chosen.

Simply identifying molecules by comparison of their EI spectra with a database search can be ambiguous. Besides, database entries are not available for every compound. Therefore, experiments were also conducted using isotope labeled glycine, namely D_2-glycine (NH_2CD_2COOH), $^{13}C_2$-glycine ($NH_2{}^{13}CH_2{}^{13}COOH$), 1-^{13}C-gly-

Chapter 6 Hydrothermal Biomass Valorization

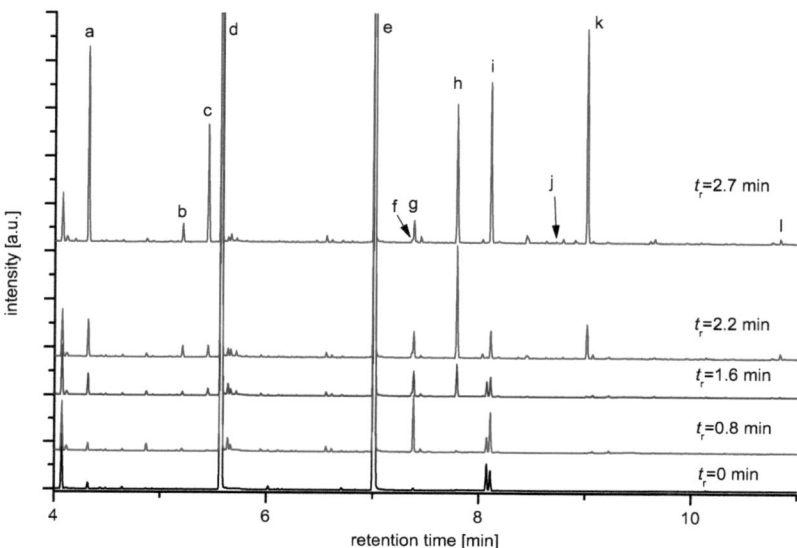

Figure 6.6: GC traces of silylated glycine decomposition products at 250 °C, reaction time as indicated.

cine ($NH_2CH_2{}^{13}COOH$), 2-^{13}C-glycine ($NH_2{}^{13}CH_2COOH$) and ^{15}N-glycine ($^{15}NH_2$ CH_2COOH). This allows the identification of the number of nitrogen and carbon atoms in an unknown compound and even gives information from which carbon center of the glycine molecule they originated. In principle the number of non-acidic hydrogens should also be accessible. However, mass spectra from a D_2-glycine sample were identical to the unlabeled compound. This proves that even non-acidic protons attached to the C_2-carbon center are easily exchanged in water.

To obtain the true number of carbon and nitrogen atoms, the molecular scaffold must stay intact and not fragmentate completely. Luckily this seems to be the case for the majority of unknown compounds. Also, when a molecular ion could not be observed, a fragment resulting from the loss of CH_3 from a TMS-group was commonly present. Most mass scans showed both the $M^{+\cdot}$ peak and a signal with $m/z = M - 15$.

One example is given in figure 6.7 for the unknown k. The mass spectra of the

6.2 Stability and Decomposition of Glycine

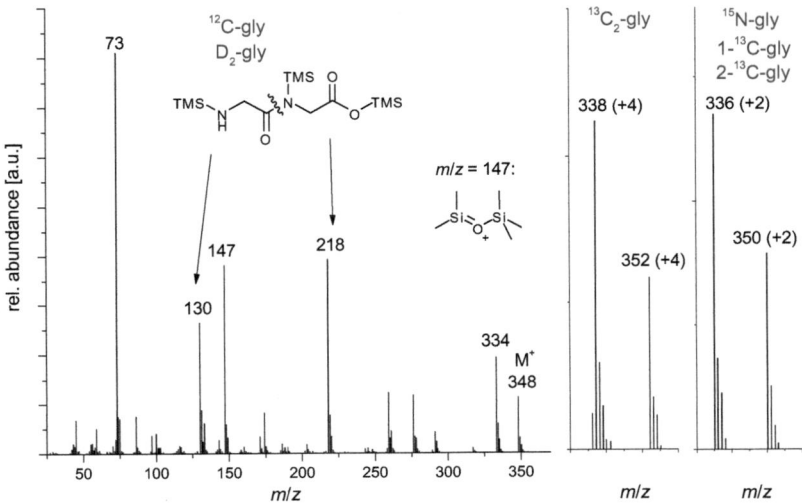

Figure 6.7: MS spectra of unknown k with non-labeled and labeled glycines.

unlabeled compound and the D_2-glycine show two peaks at high m/z-value, differing by 15 (CH_3). The peak at $m/z = 348$ is most probably the molecular ion. In the spectrum of the $^{13}C_2$-glycine both peaks are shifted by four. This reveals that the unknown compound has four carbon atoms. Both spectra of 1-^{13}C-glycine and 2-^{13}C-glycine exhibit a shift of two, which tells us that two carbon atoms from the C_1- and correspondingly two from the C_2-position of glycine are incorporated into the unknown compound. Likewise, it must contain two nitrogen atoms. Taking into account the molar mass, we can propose a trisilylated linear diglycine. This also explains the fragmentation pattern. The two main peaks $m/z = 130$ and $m/z = 218$ originate from splitting of the amide bond. These signals also show the expected shift for the isotope labeled glycines. The ions $m/z = 73$ and $m/z = 147$ do not originate from the diglycine backbone. The signal at $m/z = 73$ is the TMS$^+$ cation, which is characteristic for EI spectra of trimethylsilylated compounds. The $m/z = 147$ signal is obtained for many compounds containing at least two TMS-groups and oxygen.

It is even formed when the two silyl moieties are not vicinal [146]. The determined molecular formula can finally be verified by checking the ratio of the M, (M+1) and (M+2) peaks. It corresponds well to the calculated fractions based on the natural abundances of the different isotopes.

Other peaks were identified using the same approach. Results are summarized in table 6.1. It should be emphasized that comparing the relative peak areas of two different compounds does not yield quantitative results, because response factors can differ. Nevertheless, a rough estimation is possible, for which reason the relative peak area based on the largest signal (*tri*-TMS-glycine) is given. The table lists the masses of the ions with the largest observed m/z-value, and a probable structure of the unknown compound is proposed, based on the number of carbon and nitrogen atoms (C_xN_y). Furthermore, the number of TMS-groups (n TMS) in the derivative is given.

As the main decomposition pathway of glycine is decarboxylation, increasing concentrations of carbonic acid are found at longer reaction times (peak a). Despite its volatility, CO_2 is not lost during freeze drying, probably due to salt formation. Another product that is stabilized as its salt is N-methyl carbamic acid (peak c), which is formed from carbon dioxide and methyl amine. Despite its instability it is found in considerable amounts. This might be supported by the unique reactor design. The high pressure of 100 bar induces the formation of the carbamate. When depressurization occurs, the solution is already cooled down and was quickly quenched with liquid nitrogen and then freeze dried.

The second major reaction pathway is the dimerization to 2,5-diketopiperazine (peak h). At high temperature, condensation is obviously thermodynamically favoured. The formation of the linear dimer (peak k) does not occur before the cyclic dimer is present. Hence, diglycine seems to be formed not from the direct condensation of two glycine molecules, but from a hydrolysis of 2,5-diketopiperazine via ring opening.

Similarly, the third decomposition pathway described in literature, deamination, leading to glycolic acid (peak b) and ammonia, could be verified. This seems to be only a minor reaction. However, more compounds that have not yet been observed in the thermal decomposition of glycine could be identified, for instance amino malonic acid (peak i). Interestingly, glycine not only decarboxylates, but also incorporates CO_2 to some extent. A possible mechanism is proposed in figure 6.8. It is based on

6.2 Stability and Decomposition of Glycine

Table 6.1: Identified compounds originating from the decomposition of glycine

peak	a	b	c	d	e	f	g
rel. area	9.7%	1.0%	6.2%	31.0%	100%	0.02%	1.2%
ion	M^+-15	M^+-15	M^+-15	M^+-15	M^+-15	M^+	M^+-117
^{12}C	191	205	204	204	276	256	218
$^{13}C_2$	192	207	206	206	278	260	220
1-^{13}C	192	206	205	205	277	258	219
2-^{13}C	191	206	205	205	277	258	219
^{15}N	191	205	205	205	277	258	219
C_xN_y	C_1N_0	C_2N_0	C_2N_1	C_2N_1	C_2N_1	C_4N_2	C_2N_1
n TMS	2	2	2	2	3	2	3
structure	H_2CO_3	glycolic acid	carbamic acid	glycine	glycine	2,5-dihydroxy pyrazine	carbamic acid

peak	h	i	j	k	l
rel. area	7.4%	8.4%	0.1%	11.5%	0.3%
ion	M^+	M^+-15	M^+	M^+	M^+
^{12}C	258	320	344	348	464
$^{13}C_2$	262	323	348	352	469
1-^{13}C	260	322	346	350	467
2-^{13}C	260	321	346	350	466
^{15}N	260	321	346	350	466
C_xN_y	C_4N_2	C_3N_2	C_4N_2	C_4N_2	C_5N_2
n TMS	2	3	3	3	4
structure	2,5-diketo-piperazine	amino malonic acid	trihydroxy pyrazine	diglycine	$C_5H_8N_2O_5$

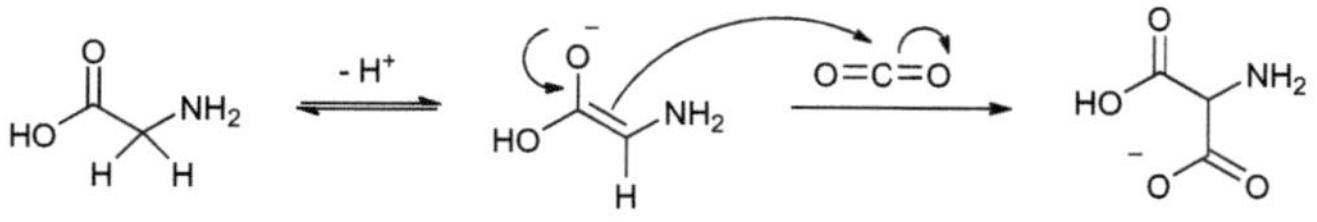

Chapter 6 Hydrothermal Biomass Valorization

Figure 6.8: Proposed mechanism for the incorporation of CO_2 into glycine.

the fact that deuterated glycine rapidly exchanges protons attached to the C_2-center. This most probably occurs via a base catalyzed abstraction of protons. The resulting nucleophilic anion can then react with the electrophilic carbon dioxide, as shown in figure 6.8. In this context it is worth noting that the ion product of water is at a maximum at 250 °C (see figure 2.7, page 25).

An interesting kinetics of formation and decomposition is observed for compound g. Its concentration is highest for short reaction time and then decreases. Analogous to N-methyl carbamic acid, this is most probably the carbamic acid of glycine. The peak with the largest observed m/z originates from the elimination of a COOTMS fragment. For short reaction time there is already enough CO_2 to form the carbamic acid, whereas for prolonged reaction times the concentration of glycine decreases and therefore so does the peak intensity. Hence, the formation of the carbamic acid is an equilibrium reaction.

Compounds f and j are aromatics that are formed when 2,5-diketopiperazine is oxidized, for instance by elimination of hydrogen. The small peak at 8.4 ppm in the ^{1}H-NMR spectrum (figure 6.5) might be due to these compounds, although formic acid also shows this chemical shift. The intensity of the chromatographic peaks f and j increased at the expense of the cyclic dimer when using higher derivatization temperatures. It is therefore not clear, whether they are an artifact from the silylation or not. In any case, the elimination of hydrogen, although not common at moderate temperature, happens more easily under hydrothermal conditions. Glycerol for instance is converted to lactic acid via this mechanism [63].

Larger compounds than diglycine can also be detected. The molar mass of peak l corresponds to a diglycine with one additional carboxylic acid group. However, the exact structure could not be determined. Larger oligomers could also be present, but they can no longer be resolved by GC. Additional HPLC measurements would be necessary.

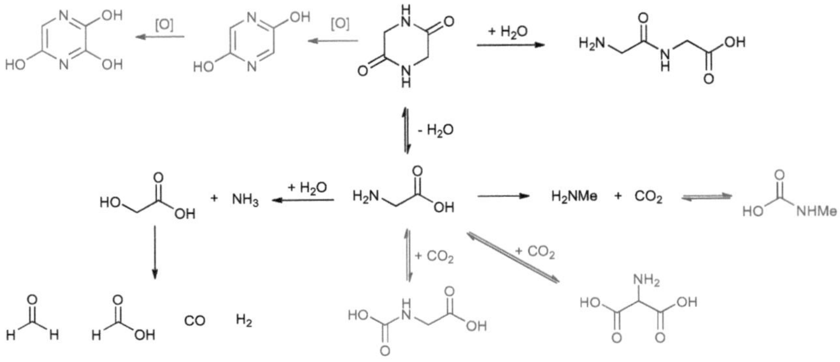

Figure 6.9: Reaction network of the hydrothermal glycine decomposition, products in gray have not yet been described.

These data provide some overview about the hydrothermal glycine decomposition. A reaction network is shown in figure 6.9, whereby products drawn in black have already been described in literature [144]. Pathways drawn in blue have not yet been observed. GC measurements thus allowed to extend the existing knowledge about the hydrothermal behaviour of glycine.

6.3 Summary

In combination with mass spectrometry, gas chromatographic measurements provided insights into the hydrothermal behaviour of glucose and glycine. Silylation turned out to be a versatile tool to derivatize and identify virtually all compounds with a molecular weight low enough to allow for gas chromatographic separation. Although this latter constraint is a limitation especially for complex biomass, some mechanistic conclusions could still be drawn for the alkaline digestion of wood. Like for the model compound glucose, the hydrothermal treatment of wood resulted in a variety of organic acids. Additional decomposition components of lignin and resin could be identified. This enabled the prediction that the hydrothermally treated solution would be feasible for anaerobic digestion, which has proven to be valid.

To clarify complex reaction networks the use of isotope-labeled compounds was ex-

Chapter 6 Hydrothermal Biomass Valorization

plored. With this technique unknown compounds could be identified. Even molecules that are instable under ambient conditions, in particular carbamic acids, could be detected. These findings allowed the extension of the already established decomposition network by further species and pathways.

Chapter 7

Conclusion and Outlook

Hydrothermal synthesis has been explored in the context of green chemistry using exemplary reactions. In particular, high value was set on using a sustainable feedstock, putting emphasis on carbohydrates as they are the main component of renewable resources. Water was used as solvent as it is environmentally benign and offers unique properties at high temperatures.

The task of the first experiments, however, was not to employ carbohydrates, but to elucidate whether they could have been formed under possible prebiotic conditions. Despite the instability that is expected for sugars, the hydrothermal synthesis was successful. It revealed some crucial differences compared to the counterpart reaction at moderate temperatures. Certain catalytically active ions were not necessary since a slightly alkaline solution was sufficient to induce reaction. Although selectivity could not be improved by high temperature, the carbohydrate composition was altered, with hexoses being present only in minor amounts, whereas there was some preference for the formation of ribose amongst pentoses.

The observation that catalysts can be avoided was successfully transferred to a second reaction. Here, carbohydrates were employed as a basis for the production of the green solvent and biofuel γ-valerolactone. The key step herein is the transfer hydrogenation of levulinic acid using formic acid, a side product of a former reaction, as hydrogen donor. Conventionally this reaction requires Ru or Pd catalysts. Under hydrothermal conditions it was possible to avoid such precious metals by exploiting the temperature dependence of acid dissociation constants. Simple sulfate was recognized as a temperature switchable base. With this additive high yield could be achieved by simultaneous prevention of waste.

In a last study complex biomass was employed instead of well defined carbohydrates. With an alkaline pretreatment it was possible to make wood accessible to anaerobic fermentation, resulting in high methane yield. The complex reaction network of glycine decomposition was also explored and clarified. Besides the products already described in literature, several new species could be identified.

To perform the hydrothermal reactions under reproducible conditions, a continuous flow setup was employed. This allowed reliable kinetic studies also for quite fast reactions and is certainly the preferred setup for large scale industrial synthesis. Gas chromatography in conjunction with mass spectrometry and isotope-labeled compounds has proven to be a powerful tool for the identification of unknowns.

In summary, these examples permit the conclusion that synthesis in high-temperature water bears a great potential in future. Many catalysts, absolutely necessary under ambient conditions, can either be completely avoided or at least replaced. Examples include certain cations or bases, as in the hydrothermal formose reaction or noble metals for hydrogenation, which could be exchanged by cheap, sustainable alternatives. In this respect water acts not only as green solvent but helps to prevent waste. Furthermore, a different reactivity has to be expected at higher temperatures. Selectivity can be shifted and unexpected reactivity may be found, for instance a fast proton exchange of even non-acidid hydrogens. The increased dissociation constant allows reactions to proceed in pure water that otherwise require strong acids or bases.

Hence, many benefits arise from using high-temperature water synthesis. However, more research is necessary to explore the full potential. Additionally, some problems have to be addressed, e.g. the increased decomposition rate at high temperature and the rarely desirable formation of side products. Our experiments have shown that it can be difficult to stabilize sensitive compounds, like carbohydrates, at high temperature. Methods applicable under ambient conditions simply do not work in HTW. This requires the development of novel techniques. In any case the shortage of fossil resources will necessitate the consideration of green alternatives, and the hydrothermal synthesis in water has already proven to be highly versatile and will certainly gain importance in the future.

Appendix

Instrumental Details

High Pressure Autoclaves

Experiments were carried out in 45 mL acid digestion vessels from Parr Instrument, equipped with a teflon inlet and a glass vial. In case of samples with high pH and where KF was added, the reaction was directly performed in the teflon inlet, as glass is etched to a great extent under these conditions at high temperature. The autoclaves were put in an oven, preheated to the desired temperature. Reaction time was controlled by a timer and the autoclaves were allowed to cool down in the oven.

Continuous Flow Reactor

The X-Cube Flash from Thales Nano was employed, operating up to 350 °C with adjustable pressure in the range of 50–180 bar. The reactor was made of Hastelloy C-22. For convenience the setup was coupled to a Gilson GX-271 autosampler. A 5 mL injection loop was filled with sample and 15 mL solution picked up after reaction to ensure complete collection.

GC-MS

A 6890N GC from Agilent was employed for quantitative and qualitative analysis. It was connected to an Agilent 5975 mass spectrometer with EI ionization (70 eV). Either an HP5-ms or a DB-225ms column were used. Type of column, temperature program and other settings were adjusted to the investigated system (details see page 43). Chromatograms were evaluated using the Agilent software ChemStation with integrated NIST 2.0 database.

HPLC

Quantitative analysis of the transfer hydrogenation of levulinic acid was performed by HPLC equipped with a refractive index detector (SCL-10AVP system controller, GT-154 degasser, LC-10ADVP pump, FCV-10AL mixing chamber and RID-10A

Appendix

detector from Shimadzu, AS-950 autosampler from Jasco). A C18 column was used for separation with an eluent consisting of 1% acetonitrile in 10 mM HCl. Citric acid was used as internal standard. Data were recorded both with a refractive index detector and an UV detector set at 220 nm.

UV/Vis-Spectroscopy

The conversion of formaldehyde was monitored photometrically using a Perkin Elmer Lambda 2 spectrometer.

NMR

NMR spectra were recorded on a Bruker DPX 400 operating at 400 MHz. Spectra were evaluated with the MestReC software. Deuterium oxide and $CDCl_3$ were used as solvents. 10% D_2O was added to aqueous samples containing volatile compounds.

Chemicals

Millipore water was used in all experiments. Formaldehyde solution was prepared by refluxing paraformaldehyde suspension until it became clear. Zinc prolate was prepared as described in [109]. Salts and solvents were purchased from various commercial suppliers. Other chemicals were obtained from the following contractors:

Sigma-Aldrich: γ-valerolactone (99%), levulinic acid (98%), citric acid (99%), BSTFA/TMCS (99:1), Pd@Al_2O_3, paraformaldehyde (95%), glycolaldehyde dimer, dihydroxyacetone dimer (97%), threitol (97%), galactose (99%), allose (98%), iditol, N-methyl imidazole (99%), sodium borohydride (97%), acetic anhydride ($\geq$ 99%), glycine (99%), labeled glycines, valine (99%), L-proline (99%).

Alfa-Aesar: erythritol (99%), arabinose (99%), 2-deoxyribose (99%), talose (97%), mannose (99%).

Acros Organics: formic acid (98%), ribose ($\geq$ 99%), chromotropic acid disodium dihydrate (p.a.).

Fluka: *myo*-inositol (p.a.), adenine ($\geq$ 99%).

Roth: xylose (99%), glucose (p.a.).

Merck: ethyl chloroformate (purum).

Table of Symbols

The page on which the meaning of the symbol is explained is added

ϵ		Dielectric constant, page 26
ρ		Density, page 25
a_i		Integrated signal area, page 43
BSTFA		Silylation agent bis-N,O-trimethylsilyl trifluoroacetamide, page 44
CI		Chemical ionization, page 42
c_i		Concentration of species i, page 43
DC		Direct current, page 42
E_a		Activation energy, page 38
ECF		Ethyl chloroformate, page 47
EI		Electron impact, page 42
f_i		Response factor, page 43
FID		Flame ionization detector, page 41
FT		Fourier transformation, page 42
GA		Glycolaldehyde, page 50
GC		Gas chromatography, page 40
HDA		High-density amorphous water, page 21
HTW		High-temperature water, page 24
IS		Internal standard, page 43
k		Rate constant of reaction, page 38
KF_i		Calibration factor, page 43
K_w		Ion product of water, page 25
LDA		Low density amorphous water, page 21
m		Mass, page 42
MSD		Mass selective detector, page 41
NCW		Near-critical water, page 24
oDM		Organic dry matter, page 93
q		Flow rate, page 40
RF		Radio frequency, page 42
RNA		Ribonucleic acid, page 49
sc		Supercritical, page 16
SCW		Supercritical water, page 23
SCWO		Supercritical water oxidation, page 23

Table of Symbols

SIM	Selective ion monitoring, page 42
TCD	Thermal conductivity detector, page 41
TMCS	Trimethylchlorosilane, page 44
TOF	Time of flight, page 42
t_r	Residence time, page 40
UCST	Upper critical solution temperature, page 26
WCOT	Well coated open tubular column, page 41
z	Charge of an ion, page 42

List of Figures

1.1	A hydrothermal vent and the striking biosphere at this environment [1].	8
2.1	The three main building blocks of lignin.	14
2.2	Distribution of energy requirements in corn-grain ethanol production, taken from [25].	15
2.3	Glucose as chiral auxiliary in a Diels-Alder reaction.	18
2.4	One water molecule can form four hydrogen bonds (calculated with *Gaussian 03*).	19
2.5	Phase diagram of water [39].	20
2.6	Molecular pair correlation functions of liquid and supercritical water at 1 kbar (a) and its first derivative (b), taken from [54].	24
2.7	Properties of water at high temperature under 250 bar, taken from [56].	25
2.8	Liquid-liquid equilibria of acetophenone (component 1)/water, taken from [30].	26
2.9	Phase diagram of NaCl/water at 250 bar, taken from [59].	27
2.10	Phase diagram of Na_2SO_4/water at 250 bar, taken from [59].	27
2.11	Hydrothermal processing regions and preferential products, taken from [25].	29
2.12	Disappearance of methyl *tert*-butyl ether; apparent rate constant (a) and pH-corrected rate constant (b), taken from [65].	31
2.13	Acid-catalyzed mechanism for the hydrolysis of 4-nitroaniline.	31
2.14	Synthesis of substituted benzimidazoles from 1,2-phenylenediamine.	32
2.15	The Beckmann rearrangement of cyclohexanoxime yields ϵ-caprolactam.	33
2.16	Scheme of synthesis reactions carried out in high-temperature water, taken from [56].	34
2.17	Conversion landscape of carbohydrates by various pathways through elimination or incorporation of H_2O and CO_2.	35
3.1	Schematic overview of a high pressure continuous flow reactor.	39
3.2	The basic setup of a GC consists of a gas supply (a), injector (b), oven and column (c), and detector (d).	41
3.3	Preparation of alditol acetates from carbohydrates.	45

List of Figures

3.4 Gas chromatogram of an alditol acetate reference mixture showing high resolution even of stereoisomers. 46
3.5 Derivatization of amino acids with ethyl chloroformate yields volatile products. 47
3.6 Proposed mechanism for the chromotropic acid assay in concentrated sulphuric acid. 48
4.1 In basic media the disproportionation of formaldehyde competes with C-C bond formation. 50
4.2 $Ca(OH)_2$ acts as base and catalyst by stabilizing the enediol form of carbohydrates, which then condensate with formaldehyde. 50
4.3 Sugar formation in a typical formose reaction (2 M formaldehyde in 0.2 M $Ca(OH)_2$ at 60 °C), taken from [104]. 51
4.4 Conversion of formaldehyde at 60 °C in a) 0.05 M $Ca(OH)_2$; b) 0.05 M $Ca(OH)_2$ with 1 mol% GA and c) 0.1 M NaOH with 1 mol% GA. ... 53
4.5 Kinetics of sugar formation at 60 °C, catalysis by $Ca(OH)_2$. 55
4.6 Conversion of formaldehyde at 200 °C and 100 bar in the presence of various salts. 56
4.7 NMR spectrum of the hydrothermal formose reaction in 0.1 M acetic acid just shows Cannizzaro products; reaction time 5.8 min at 200 °C. 57
4.8 ^{1}H-NMR spectra of a formose reaction in 0.1 M $NaHCO_3$ at 200 °C with 1 mol% dihydroxyacetone added; the lower spectrum was obtained after 0.5 min, around the yellowing point and the upper one after 2 min; spectra recorded in D_2O. 58
4.9 ^{13}C-NMR spectra of a formose reaction in 0.1 M $NaHCO_3$ at 200 °C with 1 mol% dihydroxyacetone added; the lower spectrum was obtained after 0.5 min, around the yellowing point and the upper one after 2 min; spectra recorded in D_2O. 59
4.10 GC traces of a 0.5 M formaldehyde solution in 0.1 M K_2HPO_4 reacted for 0.81 min at 200 °C and 100 bar; a) after reduction with $NaBH_4$, esterification with methanolic HCl and acetylation; b) after reduction with $NaBH_4$ and acetylation; c) directly acetylated reaction mixture. Only marginal amounts of sugar alcohols are present after reaction. . 60
4.11 Conversion of formaldehyde with addition of sodium (black curves) or calcium (gray curves) acetate in presence and absence of the initiator glycolaldehyde. 62
4.12 Kinetics of formaldehyde consumption and product formation of a 0.5 M formaldehyde solution in 0.1 M K_2HPO_4 at 200 °C. 64
4.13 GC traces of the reduced and acetylated formose products for different ions (each scaled with different factor). 65

List of Figures

4.14 Variation of pH (black squares) and calculated amount of formic acid (gray circles) for the reaction of 0.5 M formaldehyde (gray triangles) in 0.1 M K_2HPO_4 at 200 °C. 66

4.15 Kinetics of formaldehyde consumption and product formation of a 0.5 M formaldehyde solution in 0.5 M carbonate buffer at 150 °C. . . 67

4.16 Fraction of pentoses showing arabinitol (circles), ribitol (squares) and xylitol (triangles) for a) the hydrothermal reaction in 0.1 M K_2HPO_4 and b) the reaction at 60 °C in 0.05 M $Ca(OH)_2$. 70

4.17 Consumption of formaldehyde in 0.1 M K_2HPO_4 with 1 mol% GA at 200 °C (squares), 175 °C (circles), 150 °C (upward triangles) and 125 °C (downward triangles). 71

4.18 Conversion of 0.5 M formaldehyde at 200 °C in presence of borate (black curves) and borate buffer (gray curves). 72

4.19 ^{1}H-NMR of the reaction of 0.5 M formaldehyde with 0.1 M adenine at 200 °C for 5.2 min lacks of peaks corresponding to carbohydrates; recorded in D_2O. 73

4.20 ^{1}H-NMR of the precipitate from the formose reaction in the presence of adenine, reaction time 1.2 min; recorded in DMSO. 73

4.21 GC of the silylated precipitate. 74

4.22 MS of the unknown compound d in the precipitate. 74

5.1 Decomposition of glucose to levulinic and formic acid with intermediate formation of hydroxymethylfurfural. 78

5.2 Transfer hydrogenation of levulinic acid with formic acid as hydrogen donor. 78

5.3 The dependence of γ-valerolactone yield upon pH, measured before (squares) and after (circles) the experiment. 81

5.4 Amount of levulinic acid (circles), formic acid (squares) and the product γ-valerolactone (triangles) versus pH prior to reaction. 82

5.5 Variation of pK_a of HSO_4^- (a) and formic acid (b) with increasing temperature, calculated from thermodynamic data. 83

5.6 Yield versus the pK_a of the corresponding acids of the tested salts under ambient conditions and at 220 °C or 200 °C for the halides respectively. 84

5.7 Catalytic effect of Na_2SO_4 on the transfer hydrogenation. 85

5.8 Variation of yield versus reaction temperature for a residence time in the range of 18.1 min (at 175 °C) and 14.7 min (at 300 °C) with 0.125 M Na_2SO_4 added. 85

111

List of Figures

6.1 Titration of alkaline glucose digestion at 130 °C for 1 d in presence of amines; a) no additive; b) 3.7 mol% histidine hydrochloride; c) 4.9 mol% lysine; d) 11.3 mol% piperidine. 90

6.2 Chromatogram of a silylated sample from alkaline wood digestion shows typical decomposition products of carbohydrates and some fragments resulting from lignin, TMS group removed in drawings. 92

6.3 Gas yield versus fermentation time for alkaline digested wood. 93

6.4 Conversion of 1 w% glycin at 180 °C (gray circles) and 250 °C (black squares). 95

6.5 NMR of the glycine decomposition experiment at 250 °C shows residual glycine (a), cyclic dimer (b) and methylamine (c). 95

6.6 GC traces of silylated glycine decomposition products at 250 °C, reaction time as indicated. 96

6.7 MS spectra of unknown k with non-labeled and labeled glycines. . . . 97

6.8 Proposed mechanism for the incorporation of CO_2 into glycine. 100

6.9 Reaction network of the hydrothermal glycine decomposition, products in gray have not yet been described. 101

List of Tables

2.1	The Hofmeister series	23
4.1	Carbohydrate composition at reaction time with maximum yield	55
4.2	Maximum yield of different carbohydrates at hydrothermal and moderate temperatures of a 0.5 M formaldehyde solution with different salts; concentration of salt was 0.1 M except for $Ca(OH)_2$ with 0.05 M.	69
4.3	Comparison of formose reaction characteristics under hydrothermal and moderate temperatures	75
5.1	Effect of salt addition (0.5 M) on the yield of γ-valerolactone	80
6.1	Identified compounds originating from the decomposition of glycine	99

Bibliography

[1] Wikipedia. Hydrothermal vent, February 2011. http://en.wikipedia.org/wiki/Hydrothermal_vent.

[2] Paul Anastas and Nicolas Eghbali. Green chemistry: Principles and practice. *Chemical Society Reviews*, 39(1):301–312, 2010.

[3] Istvan T. Horvath and Paul T. Anastas. Innovations and green chemistry. *Chemical Reviews*, 107(6):2169–2173, June 2007.

[4] P. T. Anastas and M. M. Kirchhoff. Origins, current status, and future challenges of green chemistry. *Accounts Of Chemical Research*, 35(9):686–694, September 2002.

[5] M. Poliakoff, J. M. Fitzpatrick, T. R. Farren, and P. T. Anastas. Green chemistry: Science and politics of change. *Science*, 297(5582):807–810, August 2002.

[6] J. H. Clark. Green chemistry: challenges and opportunities. *Green Chemistry*, 1(1):1–8, February 1999.

[7] J. Bashkin, R. Rains, and M. Stern. Taking green chemistry from the laboratory to chemical plant. *Green Chemistry*, 1(2):G41–G43, April 1999.

[8] B. Kamm, P. R. Gruber, and M. Kamm, editors. *Biorefineries: Industrial Processes and Products — Status Quo and Future Directions*. Wiley, 2006.

[9] S. Fernando, S. Adhikari, C. Chandrapal, and N. Murali. Biorefineries: Current status, challenges, and future direction. *Energy & Fuels*, 20(4):1727–1737, 2006.

[10] B. Kamm and M. Kamm. Principles of biorefineries. *Applied Microbiology And Biotechnology*, 64(2):137–145, April 2004.

[11] Joseph J. Bozell and Gene R. Petersen. Technology development for the production of biobased products from biorefinery carbohydrates — the US department of energy's "Top 10" revisited. *Green Chemistry*, 12(4):539–554, April 2010.

[12] S. Varadarajan and D. J. Miller. Catalytic upgrading of fermentation-derived organic acids. *Biotechnology Progress*, 15(5):845–854, 1999.

[13] J. Lunt. Large-scale production, properties and commercial applications of polylactic acid polymers. *Polymer Degradation And Stability*, 59(1-3):145–152, 1998.

[14] N. G. Lewis and E. Yamamoto. Lignin — occurrence, biogenesis and biodegradation. *Annual Review Of Plant Physiology And Plant Molecular Biology*, 41:455–496, 1990.

[15] Mike Kleinert and Tanja Barth. Towards a lignincellulosic biorefinery: Direct one-step conversion of lignin to hydrogen-enriched biofuel. *Energy & Fuels*, 22(2):1371–1379, March 2008.

[16] V. E. Tarabanko, D. V. Petukhov, and G. E. Selyutin. New mechanism for the catalytic oxidation of lignin to vanillin. *Kinetics And Catalysis*, 45(4):569–577, July 2004.

[17] G. Scholz, J. Lohr, E. Windeisen, F. Troeger, and G. Wegener. Carbonization of hot-pressed arboform-mixtures. *European Journal Of Wood And Wood Products*, 67(3):351–355, August 2009.

[18] M. Toda, A. Takagaki, M. Okamura, J. N. Kondo, S. Hayashi, K. Domen, and M. Hara. Green chemistry — biodiesel made with sugar catalyst. *Nature*, 438(7065):178–178, 2005.

[19] Z. Helwani, M. R. Othman, N. Aziz, J. Kim, and W. J. N. Fernando. Solid heterogeneous catalysts for transesterification of triglycerides with methanol: A review. *Applied Catalysis A-General*, 363(1-2):1–10, 2009.

[20] M. A. R. Meier, J. O. Metzger, and U. S. Schubert. Plant oil renewable resources as green alternatives in polymer science. *Chemical Society Reviews*, 36:1788–1802, 2007.

[21] C. H. C. Zhou, J. N. Beltramini, Y. X. Fan, and G. Q. M. Lu. Chemoselective catalytic conversion of glycerol as a biorenewable source to valuable commodity chemicals. *Chemical Society Reviews*, 37(3):527–549, 2008.

[22] M. Pagliaro, R. Ciriminna, H. Kimura, M. Rossi, and C. Della Pina. From glycerol to value-added products. *Angewandte Chemie-International Edition*, 46(24):4434–4440, 2007.

[23] J. Barrault and F. Jerome. Design of new solid catalysts for the selective conversion of glycerol. *European Journal Of Lipid Science And Technology*, 110(9):825–830, 2008.

[24] A. J. Ragauskas, C. K. Williams, B. H. Davison, G. Britovsek, J. Cairney, C. A. Eckert, W. J. Frederick, J. P. Hallett, D. J. Leak, C. L. Liotta, J. R. Mielenz, R. Murphy, R. Templer, and T. Tschaplinski. The path forward for biofuels and biomaterials. *Science*, 311(5760):484–489, 2006.

[25] Andrew A. Peterson, Frederic Vogel, Russell P. Lachance, Morgan Froeling, Jr. Antal, Michael J., and Jefferson W. Tester. Thermochemical biofuel production in hydrothermal media: A review of sub- and supercritical water technologies. *Energy & Environmental Science*, 1(1):32–65, 2008.

[26] M. J. Earle and K. R. Seddon. Ionic liquids. Green solvents for the future. *Pure And Applied Chemistry*, 72(7):1391–1398, July 2000.

[27] L. A. Blanchard, D. Hancu, E. J. Beckman, and J. F. Brennecke. Green processing using ionic liquids and CO_2. *Nature*, 399(6731):28–29, May 1999.

[28] E. J. Beckman. Supercritical and near-critical CO_2 in green chemical synthesis and processing. *Journal Of Supercritical Fluids*, 28(2-3):121–191, March 2004.

[29] C. J. Li and L. Chen. Organic chemistry in water. *Chemical Society Reviews*, 35(1):68–82, 2006.

[30] U. M. Linsdström, editor. *Organic Reactions in Water*. Blackwell Publishing, 2007.

[31] D. C. Rideout and R. Breslow. Hydrophobic acceleration of diels-alder reactions. *Journal Of The American Chemical Society*, 102(26):7816–7817, 1980.

[32] A. Lubineau, J. Auge, and Y. Queneau. Water-promoted organic-reactions. *Synthesis-Stuttgart*, (8):741–760, 1994.

[33] S. Kobayashi. Rare-earth-metal trifluoromethanesulfonates as water-tolerant Lewis acid catalysts in organic synthesis. *Synlett*, (9):689–701, 1994.

[34] A. Corma and H. Garcia. Lewis acids: From conventional homogeneous to green homogeneous and heterogeneous catalysis. *Chemical Reviews*, 103(11):4307–4365, 2003.

[35] U. M. Lindström. Stereoselective organic reactions in water. *Chemical Reviews*, 102(8):DOI 10.1021/cr010122p|UNSP CR010122P, 2002.

[36] J. Gyarmati, C. Hajdu, Z. Dinya, K. Micskei, C. Zucchi, and G. Palyi. Asymmetric induction by amino acid ligands in chromium(II)-assisted reduction of ketones. *Journal Of Organometallic Chemistry*, 586(1):106–109, August 1999.

Bibliography

[37] A. Lubineau and Y. Queneau. Aqueous cycloadditions using glyco-organic substrates. 1. Stereochemical course of the reaction. *Journal Of Organic Chemistry*, 52(6):1001–1007, March 1987.

[38] S. W. Benson and E. D. Siebert. A simple 2-structure model for liquid water. *Journal Of The American Chemical Society*, 114(11):4269–4276, May 1992.

[39] P. W. Atkins. *Physikalische Chemie*. Wiley-VCH, 3rd edition, 2001.

[40] J. L. Finney. Water? What's so special about it? *Philosophical Transactions Of The Royal Society B-Biological Sciences*, 359(1448):1145–1163, August 2004.

[41] J. Teixeira, M. C. Bellissent-Funel, S. H. Chen, and A. J. Dianoux. Experimental determination of the nature of diffusive motions of water molecules at low temperatures. *Physical Review A*, 31(3):1913–1917, 1985.

[42] Philip Ball. Water — an enduring mystery. *Nature*, 452(7185):291–292, March 2008.

[43] G. W. Robinson, S. Zhu, Singh S., and W. E. Myron. *Water in Biology, Chemisty and Physics: Experimental Overviews and Computational Methodologies*. World Scientific Publishing, 1996.

[44] G. S. Kell. Precise representation of volume properties of water at 1 atmosphere. *Journal Of Chemical And Engineering Data*, 12(1):66–69, 1967.

[45] P. Ball. *Life's Matrix: A Biography of Water*. Farrar, Straus and Grioux, 2001.

[46] J. R. Errington and P. G. Debenedetti. Relationship between structural order and the anomalies of liquid water. *Nature*, 409(6818):318–321, January 2001.

[47] C. A. Angell. Amorphous water. *Annual Review Of Physical Chemistry*, 55:559–583, 2004.

[48] H. Kanno and K. Miyata. The location of the second critical point of water. *Chemical Physics Letters*, 422(4-6):507–512, May 2006.

[49] M. G. Cacace, E. M. Landau, and J. J. Ramsden. The Hofmeister series: salt and solvent effects on interfacial phenomena. *Quarterly Reviews Of Biophysics*, 30(3):241–277, August 1997.

[50] Y. J. Zhang and P. S. Cremer. Interactions between macromolecules and ions: the Hofmeister series. *Current Opinion In Chemical Biology*, 10(6):658–663, December 2006.

[51] P. E. Savage. Organic chemical reactions in supercritical water. *Chemical Reviews*, 99(2):603–621, 1999.

[52] P. T. Williams and J. Onwudili. Composition of products from the supercritical water gasification of glucose: A model biomass compound. *Industrial & Engineering Chemistry Research*, 44(23):8739–8749, 2005.

[53] P. E. Savage, S. Gopalan, T. I. Mizan, C. J. Martino, and E. E. Brock. Reactions at supercritical conditions - applications and fundamentals. *Aiche Journal*, 41(7):1723–1778, 1995.

[54] Y. E. Gorbaty and A. G. Kalinichev. Hydrogen-bonding in supercritical water. 1. Experimental results. *Journal Of Physical Chemistry*, 99(15):5336–5340, April 1995.

[55] N. Akiya and P. E. Savage. Roles of water for chemical reactions in high-temperature water. *Chemical Reviews*, 102(8):DOI 10.1021/cr000668w|UNSP CR000668W, 2002.

[56] A. Kruse and E. Dinjus. Hot compressed water as reaction medium and reactant — properties and synthesis reactions. *Journal Of Supercritical Fluids*, 39(3):362–380, 2007.

[57] W. L. Marshall and E. U. Franck. Ion product of water substance, 0-1000 °C, 1-10,000 bars - new international formulation and its background. *Journal Of Physical And Chemical Reference Data*, 10(2):295–304, 1981.

[58] V. Valyashko. *Hydrothermal Properties of Materials*. Wiley, 2008.

[59] M. Hodes, P. A. Marrone, G. T. Hong, K. A. Smith, and J. W. Tester. Salt precipitation and scale control in supercritical water oxidation — Part A: fundamentals and research. *Journal Of Supercritical Fluids*, 29(3):265–288, 2004.

[60] Dinesh Mohan, Jr. Pittman, Charles U., and Philip H. Steele. Pyrolysis of wood/biomass for bio-oil: A critical review. *Energy & Fuels*, 20(3):848–889, May 2006.

[61] M. Roberts, J. Williams, P. Halberstadt, D. Sanders, and T. Adams. Animal waste to marketable products. In *Natural Gas Technologies Conference*, Phoenix, Arizona, USA, November 2004.

[62] M. M. Titirici, A. Thomas, and M. Antonietti. Back in the black: hydrothermal carbonization of plant material as an efficient chemical process to treat the CO_2 problem? *New Journal Of Chemistry*, 31(6):787–789, 2007.

[63] W. Z. He, G. M. Li, L. Z. Kong, H. Wang, J. W. Huang, and J. C. Xu. Application of hydrothermal reaction in resource recovery of organic wastes. *Resources Conservation And Recycling*, 52(5):691–699, 2008.

[64] A. R. Katritzky, S. M. Allin, and M. Siskin. Aquathermolysis: Reactions of organic compounds with superheated water. *Accounts Of Chemical Research*, 29(8):399–406, 1996.

[65] J. D. Taylor, J. I. Steinfeld, and J. W. Tester. Experimental measurement of the rate of methyl tert-butyl ether hydrolysis in sub- and supercritical water. *Industrial & Engineering Chemistry Research*, 40(1):67–74, January 2001.

[66] K. Zenda and T. Funazukuri. Depolymerization of poly(ethylene terephthalate) in dilute aqueous ammonia solution under hydrothermal conditions. *Journal Of Chemical Technology And Biotechnology*, 83(10):1381–1386, 2008.

[67] J. Lu, J. S. Brown, E. C. Boughner, C. L. Liotta, and C. A. Eckert. Solvatochromic characterization of near-critical water as a benign reaction medium. *Industrial & Engineering Chemistry Research*, 41(12):UNSP IE020160E, June 2002.

[68] B. Kuhlmann, E. M. Arnett, and M. Siskin. Classical organic-reactions in pure superheated water. *Journal Of Organic Chemistry*, 59(11):3098–3101, June 1994.

[69] J. Fraga-Dubreuil and M. Poliakoff. Organic reactions in high-temperature and supercritical water. *Pure And Applied Chemistry*, 78(11):1971–1982, 2006.

[70] J. S. Brown, R. Glaser, C. L. Liotta, and C. A. Eckert. Acylation of activated aromatics without added acid catalyst. *Chemical Communications*, (14):1295–1296, 2000.

[71] A. R. Katritzky, D. A. Nichols, M. Siskin, R. Murugan, and M. Balasubramanian. Reactions in high-temperature aqueous media. *Chemical Reviews*, 101(4):837–892, 2001.

[72] P. Lidström, J. Tierney, B. Wathey, and J. Westman. Microwave assisted organic synthesis — a review. *Tetrahedron*, 57(45):9225–9283, November 2001.

[73] G. Jas and A. Kirschning. Continuous flow techniques in organic synthesis. *Chemistry-A European Journal*, 9(23):5708–5723, December 2003.

[74] M. Brivio, W. Verboom, and D. N. Reinhoudt. Miniaturized continuous flow reaction vessels: influence on chemical reactions. *Lab On A Chip*, 6(3):329–344, March 2006.

[75] M. Sasaki, K. Goto, K. Tajima, T. Adschiri, and K. Arai. Rapid and selective retro-aldol condensation of glucose to glycolaldehyde in supercritical water. *Green Chemistry*, 4(3):285–287, 2002.

[76] Brian P. Mason, Kristin E. Price, Jeremy L. Steinbacher, Andrew R. Bogdan, and D. Tyler McQuade. Greener approaches to organic synthesis using microreactor technology. *Chemical Reviews*, 107(6):2300–2318, June 2007.

[77] I. V. Perez, S. Rogak, and R. Branion. Supercritical water oxidation of phenol and 2,4-dinitrophenol. *Journal Of Supercritical Fluids*, 30(1):71–87, June 2004.

[78] E. W. Lemmon, M. O. McLinden, and D. G. Friend. *"Thermophysical Properties of Fluid System"* in NIST Chemistry WebBook, NIST Standard Reference Database Number 69. National Institute of Standards and Technology, Gaithersburg MD, 20899, http://webbook.nist.gov, retrieved October 15, 2010.

[79] G. Schwedt. *Chromatographische Trennmethoden.* Thieme, 3rd edition, 1994.

[80] Supelco Analytical. GC column selction guide, January 2011. http://www.sigmaaldrich.com/etc/medialib/docs/Supelco/General_Information/t407133.Par.0001.File.tmp/t407133.pdf.

[81] C. Dass. *Fundamentals of contemporary mass spectrometry.* Wiley, 2007.

[82] D. L. Stalling, C. W. Gehrke, and R. W. Zumwalt. A new silylation reagent for amino acids bis (trimethylsilyl) trifluoracetamide (BSTFA). *Biochemical And Biophysical Research Communications*, 31(4):616–622, 1968.

[83] G. L. Sassaki, L. M. Souza, R. V. Serrato, T. R. Cipriani, P. A. J. Gorin, and M. Iacomini. Application of acetate derivatives for gas chromatography-mass spectrometry: Novel approaches on carbohydrates, lipids and amino acids analysis. *Journal Of Chromatography A*, 1208(1-2):215–222, 2008.

[84] R. A. Laine and C. C. Sweeley. Analysis of trimethylsilyl O-methyloximes of carbohydrates by combined gas-liquid chromatography-mass spectrometry. *Analytical Biochemistry*, 43(2):533–538, 1971.

[85] A. B. Blakeney, P. J. Harris, R. J. Henry, and B. A. Stone. A simple and rapid preparation of alditol acetates for monosaccharide analysis. *Carbohydrate Research*, 113(2):291–299, 1983.

[86] T. G. Sobolevsky, A. I. Revelsky, B. Miller, V. Oriedo, E. S. Chernetsova, and I. A. Revelsky. Comparison of silylation and esterification/acylation procedures

in GC-MS analysis of amino acids. *Journal Of Separation Science*, 26(17):1474–1478, November 2003.

[87] P. Husek. Rapid derivatization and gas-chromatographic determination of amino acids. *Journal Of Chromatography*, 552(1-2):289–299, 1991.

[88] P. E. Georghiou and C. K. J. Ho. The chemistry of the chromotropic acid method for the analysis of formaldehyde. *Canadian Journal Of Chemistry-Revue Canadienne De Chimie*, 67(5):871–876, May 1989.

[89] E. Fagnani, C. B. Melios, L. Pezza, and H. R. Pezza. Chromotropic acid-formaldehyde reaction in strongly acidic media. The role of dissolved oxygen and replacement of concentrated sulphuric acid. *Talanta*, 60(1):171–176, May 2003.

[90] P. Decker, H. Schweer, and R. Pohlmann. Bioids. 10. Identification of formose sugars, presumable prebiotic metabolites, using capillary gas chromatography/gas chromatography-mass spectrometry of n-butoxime trifluoroacetates on OV-225. *Journal Of Chromatography*, 244(2):281–291, 1982.

[91] J. W. Schopf. Microfossils of the early archean apex chert — new evidence of the antiquity of life. *Science*, 260(5108):640–646, April 1993.

[92] G. Macleod, C. Mckeown, A. J. Hall, and M. J. Russell. Hydrothermal and oceanic pH conditions of possible relevance to the origin of life. *Origins Of Life And Evolution Of The Biosphere*, 24(1):19–41, February 1994.

[93] S. L. Miller. A production of amino acids under possible primitive earth conditions. *Science*, 117(3046):528–529, 1953.

[94] S. L. Miller. Production of some organic compounds under possible primitive earth conditions. *Journal Of The American Chemical Society*, 77(9):2351–2361, 1955.

[95] B. M. Rode. Peptide and the origin of life. *Peptides*, 20(6):773–786, 1999.

[96] P. Ehrenfreund, W. Irvine, L. Becker, J. Blank, J. R. Brucato, L. Colangeli, S. Derenne, D. Despois, A. Dutrey, H. Fraaije, A. Lazcano, T. Owen, and F. Robert. Astrophysical and astrochemical insights into the origin of life. *Reports On Progress In Physics*, 65(10):1427–1487, October 2002.

[97] L. E. Orgel. The origin of life — a review of facts and speculations. *Trends In Biochemical Sciences*, 23(12):491–495, December 1998.

[98] R. Shapiro. The prebiotic role of adenine — a critical analysis. *Origins Of Life And Evolution Of The Biosphere*, 25(1-3):83–98, June 1995.

[99] G. Schlesinger and S. L. Miller. Prebiotic synthesis in atmospheres containing CH_4, CO and CO_2. 2. Hydrogen cyanide, formaldehyde and ammonia. *Journal Of Molecular Evolution*, 19(5):383–390, 1983.

[100] A. Butlerow. Bildung einer zuckerartigen Substanz durch Synthese. *Justus Liebigs Annalen der Chemie*, 120:295–298, 1861.

[101] R. Larralde, M. P. Robertson, and S. L. Miller. Rates of decomposition of ribose and other sugars — implications for chemical evolution. *Proceedings Of The National Academy Of Sciences Of The United States Of America*, 92(18):8158–8160, August 1995.

[102] C. Reid and L. E. Orgel. Synthesis of sugars in potentially prebiotic conditions. *Nature*, 216(5114):455, 1967.

[103] R. Shapiro. Prebiotic ribose synthesis — a critical analysis. *Origins Of Life And Evolution Of The Biosphere*, 18(1-2):71–85, 1988.

[104] T. Mizuno and A. H. Weiss. Synthesis and utilization of formose sugars. *Advances In Carbohydrate Chemistry And Biochemistry*, 29:173–227, 1974.

[105] A. N. Simonov, O. P. Pestunova, L. G. Matvienko, and V. N. Parmon. The nature of autocatalysis in the Butlerov reaction. *Kinetics And Catalysis*, 48(2):245–254, 2007.

[106] A. G. Cairns-Smith, G. L. Walker, and P. Ingram. Formose production by minerals — possible relevance to origin of life. *Journal Of Theoretical Biology*, 35(3):601–604, 1972.

[107] A. W. Schwartz and R. M. Degraaf. The prebiotic synthesis of carbohydrates — a reassessment. *Journal Of Molecular Evolution*, 36(2):101–106, 1993.

[108] N. W. Gabel and C. Ponnamperuma. Model for origin of monosaccharides. *Nature*, 216(5114):453–455, 1967.

[109] J. Kofoed, J. L. Reymond, and T. Darbre. Prebiotic carbohydrate synthesis: zinc-proline catalyzes direct aqueous aldol reactions of α-hydroxy aldehydes and ketones. *Organic & Biomolecular Chemistry*, 3(10):1850–1855, 2005.

[110] J. B. Lambert, S. A. Gurusamy-Thangavelu, and K. B. A. Ma. The silicate-mediated formose reaction: Bottom-up synthesis of sugar silicates. *Science*, 327(5968):984–986, 2010.

[111] Sandra Pizzarello and Arthur L. Weber. Stereoselective syntheses of pentose sugars under realistic prebiotic conditions. *Origins Of Life And Evolution Of Biospheres*, 40(1):3–10, February 2010.

[112] V. A. Likholobov, A. H. Weiss, and M. M. Sakharov. Use of temperature to simplify formose sugar composition. *Reaction Kinetics And Catalysis Letters*, 8(2):155–166, 1978.

[113] Y. Nygren, S. A. Fredriksson, and B. Nilsson. Identification of sialic acid and related acids as acetylated lactones by gas chromatography mass spectrometry. *Journal Of Mass Spectrometry*, 31(3):267–274, 1996.

[114] R. N. Goldberg, N. Kishore, and R. M. Lennen. Thermodynamic quantities for the ionization reactions of buffers. *Journal Of Physical And Chemical Reference Data*, 31(2):231–370, 2002.

[115] J. L. S. Bell, D. J. Wesolowski, and D. A. Palmer. The dissociation quotients of formic-acid in sodium-chloride solutions to 200 °C. *Journal Of Solution Chemistry*, 22(2):125–136, 1993.

[116] A. Ricardo, M. A. Carrigan, A. N. Olcott, and S. A. Benner. Borate minerals stabilize ribose. *Science*, 303(5655):196–196, 2004.

[117] D. J. Lebotlan. H-1 and C-13 NMR-study of the interaction of formaldehyde on adenine and its derivatives. *Magnetic Resonance In Chemistry*, 27(3):295–298, March 1989.

[118] S. Morooka, C. Wakai, N. Matubayasi, and M. Nakahara. Hydrothermal carbon-carbon bond formation and disproportionations of C1 aldehydes: Formaldehyde and formic acid. *Journal Of Physical Chemistry A*, 109(29):6610–6619, 2005.

[119] Y. Roman-Leshkov, J. N. Chheda, and J. A. Dumesic. Phase modifiers promote efficient production of hydroxymethylfurfural from fructose. *Science*, 312(5782):1933–1937, 2006.

[120] H. B. Zhao, J. E. Holladay, H. Brown, and Z. C. Zhang. Metal chlorides in ionic liquid solvents convert sugars to 5-hydroxymethylfurfural. *Science*, 316(5831):1597–1600, 2007.

[121] Y. Takeuchi, F. M. Jin, K. Tohji, and H. Enomoto. Acid catalytic hydrothermal conversion of carbohydrate biomass into useful substances. *Journal Of Materials Science*, 43(7):2472–2475, 2008.

[122] B. Girisuta, L. P. B. M. Janssen, and H. J. Heeres. A kinetic study on the conversion of glucose to levulinic acid. *Chemical Engineering Research & Design*, 84(A5):339–349, 2006.

[123] I. T. Horvath, H. Mehdi, V. Fabos, L. Boda, and L. T. Mika. γ-Valerolactone — a sustainable liquid for energy and carbon-based chemicals. *Green Chemistry*, 10(2):238–242, 2008.

[124] H. Mehdi, V. Fabos, R. Tuba, A. Bodor, L. T. Mika, and I. T. Horvath. Integration of homogeneous and heterogeneous catalytic processes for a multi-step conversion of biomass: From sucrose to levulinic acid, γ-valerolactone, 1,4-pentanediol, 2-methyl-tetrahydrofuran, and alkanes. *Topics In Catalysis*, 48(1-4):49–54, 2008.

[125] H.S. Broadbent and T.G. Selin. Rhenium catalysts. VI. Rhenium(IV) oxide hydrate. *Journal Of Organic Chemistry*, 28:2343–2345, 1963.

[126] P. Maki-Arvela, J. Hajek, T. Salmi, and D. Y. Murzin. Chemoselective hydrogenation of carbonyl compounds over heterogeneous catalysts. *Applied Catalysis A-General*, 292:1–49, 2005.

[127] Z. P. Yan, L. Lin, and S. J. Liu. Synthesis of γ-valerolactone by hydrogenation of biomass-derived levulinic acid over Ru/C catalyst. *Energy & Fuels*, 23(8):3853–3858, 2009.

[128] H. Heeres, R. Handana, D. Chunai, C. B. Rasrendra, B. Girisuta, and H. J. Heeres. Combined dehydration/(transfer)-hydrogenation of C6-sugars (D-glucose and D-fructose) to γ-valerolactone using ruthenium catalysts. *Green Chemistry*, 11(8):1247–1255, 2009.

[129] L. Deng, J. Li, D. M. Lai, Y. Fu, and Q. X. Guo. Catalytic conversion of biomass-derived carbohydrates into γ-valerolactone without using an external H_2 supply. *Angewandte Chemie-International Edition*, 48(35):6529–6532, 2009.

[130] T. A. Bryson, J. M. Jennings, and J. M. Gibson. A green and selective reduction of aldehydes. *Tetrahedron Letters*, 41(19):3523–3526, 2000.

[131] Daniel Kopetzki and Markus Antonietti. Transfer hydrogenation of levulinic acid under hydrothermal conditions catalyzed by sulfate as a temperature-switchable base. *Green Chemistry*, 12(4):656–660, April 2010.

[132] M. Siskin and A. R. Katritzky. Reactivity of organic compounds in superheated water: General background. *Chemical Reviews*, 101(4):825–835, 2001.

[133] M. H. Kim, C. S. Kim, H. W. Lee, and K. Kim. Temperature dependence of dissociation constants for formic acid and 2,6-dinitrophenol in aqueous solutions up to 175 °C. *Journal Of The Chemical Society-Faraday Transactions*, 92(24):4951–4956, 1996.

[134] A. G. Dickson, D. J. Wesolowski, D. A. Palmer, and R. E. Mesmer. Dissociation-constant of bisulfate ion in aqueous sodium-chloride solutions to 250 °C. *Journal Of Physical Chemistry*, 94(20):7978–7985, 1990.

[135] B. N. Ryzhenko and O. V. Bryzgalin. Dissociation of acids under hydrothermal conditions. *Geokhimiya*, (1):137–142, 1987.

[136] R. A. Bourne, J. G. Stevens, J. Ke, and M. Poliakoff. Maximising opportunities in supercritical chemistry: the continuous conversion of levulinic acid to γ-valerolactone in CO_2. *Chemical Communications*, pages 4632–4634, 2007.

[137] G. C. A. Luijkx, F. van Rantwijk, H. van Bekkum, and M. J. Antal. The role of deoxyhexonic acids in the hydrothermal decarboxylation of carbohydrates. *Carbohydrate Research*, 272(2):191–202, 1995.

[138] L. Hough, J. K. N. Jones, and E. L. Richards. The reaction of amino-compounds with sugars. 1. The action of ammonia on D-glucose. *Journal Of The Chemical Society*, (OCT):3854–3857, 1952.

[139] Z. Srokol, A. G. Bouche, A. van Estrik, R. C. J. Strik, T. Maschmeyer, and J. A. Peters. Hydrothermal upgrading of biomass to biofuel; studies on some monosaccharide model compounds. *Carbohydrate Research*, 339(10):1717–1726, July 2004.

[140] Renewable Energy Concepts. Bioenergie aus Biogas, February 2011. http://www.renewable-energy-concepts.com/german/bioenergie.html.

[141] T. Willner, P. Scherer, D. Meier, and W. Vanselow. Vergärung von Flash-Pyrolyseöl aus Holz zu Biogas. *Chemie Ingenieur Technik*, 76(6):838–842, 2004.

[142] D. K. Alargov, S. Deguchi, K. Tsujii, and K. Horikoshi. Reaction behaviors of glycine under super- and subcritical water conditions. *Origins Of Life And Evolution Of The Biosphere*, 32(1):1–12, February 2002.

[143] E. Imai, H. Honda, K. Hatori, A. Brack, and K. Matsuno. Elongation of oligopeptides in a simulated submarine hydrothermal system. *Science*, 283(5403):831–833, February 1999.

[144] D. Klingler, J. Berg, and H. Vogel. Hydrothermal reactions of alanine and glycine in sub- and supercritical water. *Journal Of Supercritical Fluids*, 43:112–119, 2007.

[145] J. L. Bada, S. L. Miller, and M. X. Zhao. The stability of amino-acids at submarine hydrothermal vent temperatures. *Origins Of Life And Evolution Of The Biosphere*, 25(1-3):111–118, June 1995.

[146] J. A. McCloskey, R. N. Stillwell, and A. M. Lawson. Use of deuterium-labeled trimethylsilyl derivatives in mass spectrometry. *Analytical Chemistry*, 40(1):233–236, 1968.

Die VDM Verlagsservicegesellschaft sucht für wissenschaftliche Verlage abgeschlossene und herausragende

Dissertationen, Habilitationen, Diplomarbeiten, Master Theses, Magisterarbeiten usw.

für die kostenlose Publikation als Fachbuch.

Sie verfügen über eine Arbeit, die hohen inhaltlichen und formalen Ansprüchen genügt, und haben Interesse an einer honorarvergüteten Publikation?

Dann senden Sie bitte erste Informationen über sich und Ihre Arbeit per Email an *info@vdm-vsg.de*.

Sie erhalten kurzfristig unser Feedback!

VDM Verlagsservicegesellschaft mbH
Dudweiler Landstr. 99
D - 66123 Saarbrücken

Telefon +49 681 3720 174
Fax +49 681 3720 1749

www.vdm-vsg.de

Die VDM Verlagsservicegesellschaft mbH vertritt

Printed by Books on Demand GmbH, Norderstedt / Germany